# SpringerBriefs in Environmental Science

More information about this series at http://www.springer.com/series/8868

Frederic R. Siegel

# Mitigation of Dangers from Natural and Anthropogenic Hazards

## Prediction, Prevention, and Preparedness

 Springer

Frederic R. Siegel
Emeritus Professor
The George Washington University
Washington, DC, USA

ISSN 2191-5547                ISSN 2191-5555  (electronic)
SpringerBriefs in Environmental Science
ISBN 978-3-319-38874-8       ISBN 978-3-319-38875-5  (eBook)
DOI 10.1007/978-3-319-38875-5

Library of Congress Control Number: 2016940909

Printed on acid-free paper

This Springer imprint is published by Springer Nature
The registered company is Springer International Publishing AG Switzerland

*I dedicate this book to my wife, Felisa,*
*who encourages my writings as much today*
*as she did in past years; to Naomi, Coby,*
*and Noa Benveniste and Solomon*
*and Beatrice Gold, my grandchildren;*
*and to the children of their generation*
*worldwide. I hope that now and in the future*
*they will benefit from governments*
*and planners that adopt applicable measures*
*discussed in the book, as well as others*
*that will be developed in the future, to lessen*
*the impacts of the many and varied hazard*
*events that can disrupt societies. I thank*
*the many colleagues whose presentations*
*I have heard and journal papers and books*
*I have read over the years that guided me*
*in writing the book.*

# Preface

On September 18, 2014, the World Health Organization (WHO) reported that the natural hazard ebola epidemic affecting Liberia, Sierra Leone, and Guinea in West Africa had caused 2630 deaths from more than 5357 persons infected by the ebola virus. By July 26, 2015, close to 2 years after the outbreak, at least 27,748 people had been infected with more than 11,279 dead (~40 % mortality rate). That a hazard can tear at the very fabric of societies was evident in West Africa. Many citizens did not go to work for fear of becoming infected with the ebola virus. This caused a slowing or stoppage of commerce and industrial output, disruptions in transport (e.g., bus and taxi service), and a closure of border crossings, airports, and seaports to traffic from countries where the ebola epidemic was not yet controlled in order to prevent entry of citizens that could be infected with the virus that takes 2–21 days to incubate. With no new cases after 21 days, the epidemic can be considered over. At the beginning of March 2015, no new cases had been reported in Liberia, and after a 42-day countdown, Liberia was declared ebola-free. In October 2015, Guinea began the 42-day countdown to being declared free of the disease. However, in November 2015, three new cases of ebola were diagnosed, and health professionals are working to find their source. By the end of 2015, West Africa was declared free of the disease. Other examples of the impacts of twenty-first-century hazards on societies' normal day-to-day activities and what we learned from each disaster that can help us to mitigate the effects of future like events are cited in this book (e.g., in Haiti [earthquake], Japan [tsunami, radiation from damaged nuclear power facility], Indonesia [tsunami], Australia [drought, wildfires], China [flooding, heavy metal health problems], India [flooding], Iceland [volcanic ash eruptions], Europe, [heat waves, flooding], and hazards in other countries). Table 1.1 gives examples of natural and anthropogenic-intensified hazards and the deaths they caused in various parts of the world since the beginning of the twenty-first century to 2015 [1].

Each year sees natural and anthropogenic hazards and secondary events triggered by them, or combinations thereof, threaten, sicken, injure, and kill people globally and damage and destroy buildings and infrastructure. Some are hazards that can last for a year or longer (e.g., drought, epidemics/pandemics), some are instantaneous or last for short periods of time (e.g., earthquakes, tsunamis, mudslides), and some may last for days, weeks, or months (e.g., floods, heat waves, volcanic eruptions) or recur at somewhat regular intervals (floods). Still others are continuous health threats in given climates such as mosquito-borne malaria, dengue fever, and Zika that have no preventive vaccines and some diseases that have preventive vaccines that are not applied to all (e.g., measles, polio) because of religious belief, coercion by religious zealots, or published misinformation. This endangers the health of the general population as indicated by the measles outbreak in the United States in 2015. Some settlements that were not thought to be sites vulnerable to a hazard when originally established are at risk today. This is because of inadequate land-use planning at the time because planners did not have the benefit of the knowledge about hazards and conditions that favor them that we have today (e.g., floods, pollution, subsidence). In addition, increasing global populations and population density, plus global warming/climate change, were not major actors that affected hazard risk and land-use planning until the second half of the twentieth century.

As the knowledge bases for different natural and anthropogenic hazards have grown over time after analysis of each new event that occurs, so has our ability to mitigate their impacts on society. Hazard mitigation can be achieved in some cases by using prediction and protection methods coupled to early warning systems (e.g., dams and levees for floods), by prevention in other cases with technological advances (e.g., improved building codes and materials, pollutant capture and control systems), by up-to-date land-use planning and preparedness, and by other technological, political, social, and economic means. Control or elimination of diseases and epidemics that may develop can be treated by biotechnical advances in medications and vaccines and by protocols to deal with them. These are discussed in the text that follows.

Washington, DC, USA                                                          Frederic R. Siegel
March, 2016

# Contents

# Chapter 1
# Introduction to Hazards and Factors that Affect Impacts on Global Societies

Hazards are sources of danger for humans and other life forms, the ecosystems they inhabit, peoples' homes and workplaces, and infrastructure that supports citizens daily activities.

Hazard events may be natural (e.g., earthquakes), may be purely the result of human actions (e.g., heavy metal pollution), or may be natural but heightened by human actions or decisions (e.g., extreme weather conditions). For some events, there is no control of exactly where they will strike (drought), how often they can be expected to occur, or for how long they will last. For others, we know where the dangers are (active earthquake fault zones, volcanoes) but there is no infallible prediction of the timing or frequency of a happening. For still others (e.g., floods after torrential or sustained rains), we can monitor selected parameters (e.g., volume and flow rate of drainage basin waterways) and predict where, when, and with what intensity the hazard will impact people and their environments. Primary hazards such as cited above can trigger secondary events that can be more destructive to life and property (e.g., earthquake caused tsunamis, landslides, and mudflows/debris flows) that in some cases may be monitored and warnings to evacuate given (e.g., tsunamis). However, in other cases, events happen suddenly with deadly results (e.g., mud flows/debris flows). There are observations, measurements, land use planning norms, technological advances, and preparedness steps that if accessible and managed correctly can mitigate and perhaps eliminate the dangers to people from a spectrum of natural and anthropogenic hazard events.

Infrastructure supports human habitation on our planet and can be considered as having multiple elements for its security that must be planned for and protected so as to minimize the impact of a hazard/disaster and support the infrastructural functionality. This being the aim, infrastructure elements should be prioritized in terms of their critical level for population protection purposes, risk management assessments, and proactive preparedness planning [1]. This book categorizes four infrastructure elements. One includes roads, bridges, railroads, airports, navigable inland waterways, and seaports, all of which must be functioning to maintain commerce in the delivery of goods and services. A second is the secure delivery and control of the

© Springer International Publishing Switzerland 2016
F.R. Siegel, *Mitigation of Dangers from Natural and Anthropogenic Hazards*,
SpringerBriefs in Environmental Science, DOI 10.1007/978-3-319-38875-5_1

utilities water, electricity, and natural gas that allow populations to go about their daily routines. Another is related to delivery of potable water and food to sustain populations as well as access to shelter, adequate sanitation services, healthcare, and education that a society needs to operate efficiently and effectively. A fourth is secure operation of communication services, warning/alert systems, search and rescue capability, and security services. If one or more of these is damaged beyond ready repair and no backup is immediately available, recovery from a hazard/disaster would be slowed.

There are multiple factors that influence where, why, and when natural or anthropogenic disasters strike vulnerable populations causing injuries, death and destruction of property, or expose citizens to diseases that can evolve into life-threatening, epidemics. These include geologic environments (tectonics and rock/soil types), geographic settings (coastal and inland zones), and topography (highlands, lowlands, valleys, plains-landforms). In addition to these mainly physical conditions, meteorology plays a major role in hazard events and their impacts on people (climatic zones, storm tracks, wind direction—global warming/climate change). Other factors that affect the consequences of hazard events include land use planning, industrial locations, population density, available and accessible medical services, societal safety nets, cultural norms, government decisions, and a government's economic strength. For some hazards (e.g., floods), prediction to a reasonable degree of when an event will happen is possible giving citizens time to prepare and evacuate, and for municipalities to ready safe sites for evacuees. Prevention is achievable for some diseases that put citizens at risk of sickness and death (e.g., by vaccination, medications). Mitigation is possible against many sources of danger to populations thus providing some protection for citizens and in some cases their property. However, mitigation requires economic investment in equipment and human resources as well as municipal and governmental planning (e.g., flood control, tsunami warning system). Investment can be worthwhile because natural disasters can slow economic growth from a national perspective and certainly can cause unemployment, social stress, and a loss of tax income. One study evaluated flooding in 187 countries over a 50 year period to 2010 and found that if the total number of people hurt by flooding was 0.1 % of the population, the per capita growth rate decreased by 0.005 %, not a huge percentage but one that can be meaningful for less developed and developing countries in terms of their economic growth [2]. Where a major earthquake, perhaps with triggered hazards, impacts a country killing hundreds or thousands, injuring more, and damaging or destroying buildings and infrastructure, the decrease in economic growth can be expected to be high such as in Haiti (earthquake, 2010). Losses here exceeded the country's gross domestic product [GDP] for a few years but recovered to 3.8 % by 2014. In Japan (earthquake, tsunami, rupture of nuclear power facility, Fukushima, 2011), GDP dropped 3.7 % for first quarter of 2011 and in 2014 was slow at 1.3 %, as the economic effects of the complex disaster were still being felt 4 years later. Prevention and mitigation are realistic options for some hazards if the "why" a hazard occurs is understood so that measures can be implemented to damp its effects and protect lives and property.

There is the academic question of when a natural hazard becomes a disaster. *Natural hazards* are regularly happenings worldwide but may or may not be a threat to the well being of people, their living environments and daily routines. This depends on where they occur, whether in populated regions such as metropolitan urban centers or lightly populated (by humans) environments such as the Sahara or Gobi deserts, Siberia, or Alaska's Aleutian islands. For example, an 8.1 magnitude earthquake struck the Gobi desert region in December, 1957 but caused only 30 deaths and may not be categorized as a disaster by some but considered so by others. Earthquakes and/or their triggered hazards become *natural disasters* when many people are injured or killed, buildings and infrastructure damaged or destroyed, and there is disruption of human activities such as happened in northeast Japan in March, 2011. Here, a magnitude 9.0 earthquake offshore and a fast moving mega-tsunami it caused impacted the region killing almost 16,000 people, displacing about 230,000 more whose homes and places of business were destroyed, as was a nuclear power facility with the release of high level radioactive pollutants. Table 1.1 presents examples of disasters during this century where deaths vary greatly depending on the hazard that caused the disaster, where and when it occurred, and any event the hazard triggered. The same distinction of natural haz-

**Table 1.1** Examples of disasters during this century where deaths vary greatly depending on the type of disaster and where it occurs [10]

| Year | Month | Region | Disaster | Deaths (estimate) |
| --- | --- | --- | --- | --- |
| 2003 | June–August | Europe | Heatwave | 35,000–70,000 |
| 2003 | December | Bam, Iran | Earthquake | 26,271 |
| 2004 | December | Asia (esp. Sumatra) | Earthquake + Tsunami | 230,000–280,000 |
| 2005 | August | Louisiana, USA | Hurricane | 1836 |
| 2005 | October | Kashmir | Earthquake | 74,500 |
| 2008 | May | Myanmar | Cyclone + storm surge | 138,366 |
| 2008 | May | Sichuan, China | Earthquake | 87,587 |
| 2008 | February | Afghanistan | Blizzard | 926 |
| 2008 | January | China | Winter storms | 133 |
| 2008 | January | Brazil | Mudslides + Floods | 128 |
| 2009 | March–December | Global | Swine Flu | 11,690 |
| 2009 | February–March | Australia | Wildfires | 173 |
| 2010 | January | Haiti | Earthquake | 160,000 |
| 2010 | June–August | Russia | Heatwave | 15,000–56,000 |
| 2011 | January | Brazil | Mudslides + Floods | 903 |
| 2011 | March | Japan | Earthquake + Tsunami | 15,884 |
| 2011 | May | Missouri, USA | Tornados | 160 |
| 2011 | April | Southeast USA | Tornados | 336 |
| 2012 | January–March | Afghanistan | Avalanches | 201 |
| 2013 | November | Philippines | Hurricane | 6201 |
| 2015 | February | Afghanistan | Avalanches | 310 |
| 2015 | June | China | Tornado | 442 |

ards to disasters outcomes in sparsely populated vs. populated urban centers is applied to volcanic activity, floods, landslides, hazards in arid and semi-arid environments, and other natural hazards that happen on our planet [3]. In contrast, *anthropogenic hazards* most often are disastrous to humans and other life forms. Such disasters affect human health and physical security either as short term events (e.g., flooding, landslides) or delayed and often long term (e.g., pollution [air, water, soil], sickness, and infectious disease).

The aims of this book are to explore established methods that can predict the onset of natural or anthropogenic hazards with reasonable lead time and also to investigate conceptual ones that may allow us to do so in the future. The book evaluates the possibility of preventing hazards, and thus protect the earth's growing populations and densely populated urban centers from harmful, dangerous events. Lastly, the text reviews technological, socio-political, and economic methods on how to lessen a hazard's impact on populations and on ecosystems that provide people with the essences of life…water and food and other natural resources. This is done for the 2015 population of 7.3 billion earthlings and for projected populations of 9.8 billion people in 2050 and perhaps 11 billion or more in 2100 [4]. A valid concern of regional and international organizations charged with designing plans to sustain future inhabitants is whether the earth can provide for increased populations. This in light of the facts that in 2015, there are at least 795 million people known to be suffering chronic malnutrition, almost an equal number without access to safe water, and 2.5 billion without access to adequate sanitation [5, 6]. For example, with respect to availability of food, the Food And Agriculture Organization of the United Nations reports the production of cereals (e.g., maize [corn], rice, wheat, barley, sorghum, and millet) in 2014 and 2015 was >2.5 billion tonnes [7]. In theory, these cereals produce enough calories alone (>2800 cal per capita per day) to nourish the projected 11 billion or more humans on earth at the end of this twenty-first century. In reality, however, 55 % of the cereals production (>1.3 tonnes) is used as feed for livestock, for conversion to biofuels, for seed and other uses, and some is wasted [8]. This, plus growing populations, especially in developing countries, their increasing rate of draw on natural resources, and the disruptions of ecosystems that put thousands of non-human species in the endangered group and at risk of extinction, signals planners that population rise may be unsustainable [9].

We will consider whether these existing problems to sustain populations can be solved and if so, how? Urban centers are a special problem with respect to hazard risk/vulnerability and disaster reduction programs. In 2015, there were 34 megacities (>10 million inhabitants) in the world [10]. Of these, 12 had population greater than 20 million people with nine of these in Asia (Table 1.2). Asia has an expected population growth from 4.4 billion in 2015 to 4.9 billion people by 2035. In Africa, the population will increase from 1.2 billion people to 1.7 billion during that period, mostly in Sub-Saharan Africa where the growth will rise from close to one billion people to 1.4 billion. In 2014, urban centers housed 53 % of the world population (~3.7 billion versus 3.6 billion in rural areas). They are projected to have ~70 % of the global population by 2050 (6.9 urbanites versus 2.9 billion rural

**Table 1.2** Mega-cities (agglomerations) with populations greater than 20 million people [11]

| Mega-city | Population $\times 10^6$ |
|---|---|
| Tokyo-Yokohama | 37.8 |
| Jakarta | 30.5 |
| Delhi, DL-UP-HR | 25.0 |
| Manila | 24.1 |
| Seoul-Incheon | 23.5 |
| Shanghai, SHG-JS-ZJ | 23.4 |
| Karachi | 22.1 |
| Beijing, BJ | 21.0 |
| New York, NY-NJ-CT | 20.6 |
| Guangzhou-Foshan, GD | 20.6 |
| Sao Paulo | 20.4 |
| Mexico City | 20.0 |

residents), mainly in Asia and Africa. In 2015, Asia had a 47% urban population, Africa 40%, South America 84%, and the developed/industrialized nations 77% [4]. We can project what could happen if the problems of chronic malnutrition, lack of access to safe water, and poor access to adequate sanitation are allowed to fester and continue to grow as global populations increase and are exposed to more energetic, more frequent, and longer lasting hazards? The text will examine what will be necessary to resolve such problems in economic terms and whether nations have the political will to do so? Clearly, careful land-use planning with respect to the present and the future, a strengthened physical infrastructure, and the population size and density are principal factors that influence the impact of hazards on urban agglomerations. If the social and economic problems are not solved, can we expect major population crashes that will directly affect the entire global community?

# References

1. Fekete, A. (2012). Common criteria for the assessment of critical infrastructures. *International Journal of Disaster Risk Science, 2*, 15–24.
2. Shabnam, N. (2014). Natural disasters and economic growth: a review. *International Journal of Disaster Risk Science, 5*, 157–163.
3. Burton, I., Kates, R. W., & White, G. F. (1978). *The environment as hazard* (258 pp.). New York: Oxford University Press. 2nd edition, 1993, The Guilford Press, 284 pp.
4. Population Reference Bureau. (2015). *World population data sheet* (21 pp.). Washington, DC: Population Reference Bureau.
5. Food and Agricultural Organization of the United Nations (FAO) (2015). *The state of food security in the world 2015* (56 pp.). Rome: FAO.

6. World Health Organization, & UNICEF. (2014). *Progress on drinking water and sanitation*, 2014 Update (75 pp.). Geneva: World Health Organization; New York: UNICEF.
7. Food and Agricultural Organization of the United Nations (FAO). (2016). World food situation, cereal supply and demand brief. http://www.fao.org/worldfoodsituation/csdb/en/.
8. Food and Agricultural Organization of the United Nations (FAO). (2015). *FAO statistical yearbook 2013. World food and agriculture* (289 pp.). Rome: FAO.
9. White, R. E. (2016). No myth: population rise unsustainable. *Nature, 589,* 283. doi:10.1038/529283c.
10. Wikipedia (accessed 2015). List of natural disasters by death toll. Table 1.1 compiled from this source. Access at: http://en.wikipedia.org/wiki/List_of_natural_disaster_by_death_toll.
11. Demographia. (2015). *Demographia world urban atlas*, 11th annual edition, 2015: 1 (134 pp.). http://www.demographia.com/db-worldua-pdf.

# Chapter 2
# Mitigation

## 2.1 Mitigation Defined

Mitigation of the impacts of natural and anthropogenic hazards (e.g., earthquakes and pollution, respectively) and the secondary hazards they may trigger (e.g., tsunamis, health problems) can be categorized as a four sequence program starting with preparedness, followed then by recovery response, recuperation/rehabilitation, and reconstruction. Each sequence can have substages carefully planned and technologically up-to-date with the aim, first of minimizing injuries and death that affected citizenry, and second to stabilize social fabrics by taking into consideration factors that can reduce physical, medical, and mental health caused stress on populations. This is done, first by supplying sufficient water, food, shelter and sanitation facilities for displaced populations (to minimize the chances of sickness and disease or their spread), and by having healthcare readily available. This may be followed by protecting property, and infrastructure to maintain services such as transportation, water, gas, and electricity. This is accomplished by preplanning to reduce the risk from a disaster (e.g., building structures according to code, flood monitoring and warning systems) and by sustaining educational and employment opportunities, thus maintaining a functioning economy.

## 2.2 Preparedness

Disaster preparedness in mitigation scenarios originates with a local government and may evolve into a national effort, or in a dire situation may require an international response. There are several elements considered essential to preparedness programs. These include the following: (1) establish a coordinated leadership; (2) identification of one or more hazards likely to impact a location (site vulnerability); (3) increase public awareness through education and training such as practiced in

© Springer International Publishing Switzerland 2016
F.R. Siegel, *Mitigation of Dangers from Natural and Anthropogenic Hazards*,
SpringerBriefs in Environmental Science, DOI 10.1007/978-3-319-38875-5_2

Japan and California USA in response to earthquakes; (4) emphasize what citizens should do when alerted to the probable onset of a hazard event (e.g., flooding) or the sudden onset of a hazard (e.g., earthquake) including evacuation plans to safe shelters stocked with water, food, and medical supplies; (5) a risk assessment for potential disasters; (6) investment in applied research and technology transfer of successful new mitigation methods; (7) incentives to apply new preventive measures that can lessen an impact of a hazard (e.g., flood control methods); and (8) assure the readiness of medical and search and rescue teams to respond. Experts in the subject emphasize that a disaster preparedness and management program can be improved in a community, large or small, urban or rural, by education of citizens and higher education that prepares cadre responding to a hazard to be flexible and able to adapt to social and cultural norms in a society [1]. They further believe that a system for enhanced information flow is necessary for locales to be able to link to a global network geared to disaster risk management. Impacted locales can communicate via computer the needs they have or will have as the result of a hazard and that may be served quickly by transportation mobility capability the world now has. The global networking during the West African Ebola outbreak was responsible for rapid response by governments, businesses and the people themselves, notwithstanding the WHO 5 month delay in declaring a public emergency. This response helped rein in the epidemic by bringing in medical experts and supplies, setting up clinics to care for the afflicted, educating the public on the transmission of the disease and how to prevent its spread, and reestablishing critical infrastructure. As noted in the Introduction, 2 years after the Ebola outbreak, West Africa was free of the Ebola disease.

Medical response to disasters depends on the infrastructure an impacted locale has (e.g., health facilities, transportation routes), trained personnel on call (e.g., medical doctors, nurses, search and recovery personnel), and supplies (e.g., medicines, vaccines). When a natural or anthropogenic hazard brings disaster to a location, hospital or health clinic personnel go into a triage mode and identify patients with the most serious conditions that require immediate treatment. Outpatient treatment should include counseling for Post Traumatic Stress Disorder.

It is especially important to regularly update hazard response protocols so as to be able to meliorate chaotic situations that can develop during the first few days after a disaster impacts a location. The preparedness extends into post-hazards planning for recuperation, rehabilitation, and reconstruction with the purpose of reducing the degree of vulnerability (exposure) of a location from like future events. This requires risk assessment and its regular updates.

# Reference

1. Jensen, S. J., Feldmann-Jensen, S., Johnston, D. M., & Brown, N. A. (2015). The emergence of a globalized system for disaster risk management and challenges for appropriate governance. *International Journal of Disaster Risk Science, 6*, 87–93.

# Chapter 3
# Risk Assessment/Vulnerability

## 3.1 Risk Assessment Analysis

Risk assessment analysis is determined for specific locations/regions with a history of being targeted by one or more primary hazards or triggered events, the probability of recurrence, the magnitude, the frequency of occurrence in a damaging/destructive magnitude range, and the projected estimates of injury, deaths, and damage and destruction. What must be also be factored in to risk assessments are the control of infectious disease and diarrhea illnesses, and mental health needs. Risk assessment analysis establishes the socio-economic value of risk related to a specific hazard (and triggered events), or injury or death in a human population and other organisms, and damage to the environment if a hazard and its effects cannot be lessened. Risk can be evaluated in a multi-step process. The first step is to identify hazard(s) and triggered destructive events they may generate and to map their intensity and the extent of their reach. A second is to catalogue the human and economic assets to be protected in an area. The third is to estimate the potential losses of life, magnitude of personal injury, damage to an economy (property loss, loss of employment, loss of tax revenue), and loss of social services expected from a hazard. Risk assessors estimate financial costs to restore conditions that can better withstand a disaster, and/or medical/death payouts to both the insured and uninsured (e.g., government grants, no cost or low cost loans). The fourth step is to determine the probability that a hazard with a specific magnitude range will occur in a given time frame.

© Springer International Publishing Switzerland 2016　　　　　　　　　　　　　9
F.R. Siegel, *Mitigation of Dangers from Natural and Anthropogenic Hazards*,
SpringerBriefs in Environmental Science, DOI 10.1007/978-3-319-38875-5_3

## 3.2   Role of Public Health

There is no doubt that a prepared and capable public health service has as its mission to prevent disease, promote health, and prolong life. This can have an important role in lessening the impact of natural and anthropogenic disasters and should be important in risk assessment analysis. A well trained and supplied public health service will treat physical injuries and deal with infectious and diarrhea diseases that can originate from disasters, and will also provide for post-disaster treatment for mental health conditions distressed populations can suffer. However, many less developed and developing nations have not invested in public health services that can respond effectively to disasters and secure societal protection from negative impacts during and post disaster [1]. A lack of full public health services impedes government planning for urban development and population growth and will manifest itself in an inadequate response for societal protection and well being during and after disasters. This dictates that there be investment in health care planning and capacity building in these countries [2]. This means having hospitals and clinics in place with well trained medical personnel and support staff, with medicines and other necessary supplies, with ambulances to transport injured and sick persons to health facilities, and with backup generators secured against physical and water threats to them if a hazard damages a disaster site normal electric grid [1]. In addition, there should be ready communication with health professionals globally and access to WHO fact sheets that describe protocols for treatment of diseases often associated with natural disasters, including vaccinations for disease control, post-disaster planning for treatment of people with chronic diseases, and access to mental health professionals [3]. Availability and access to health facilities should be equal for vulnerable groups and economically advantaged and politically connected groups.

## 3.3   Risk Reduction Funding Shortfall

The lack of investment to implement disaster risk reduction, can cost far more than expenditures to repair damages and restore what was destroyed. Investment cannot compensate for lost lives and injury. Spending on disaster risk reduction globally was $13.6 billion from 1991 through 2010. However, about $68 billion was spent on emergency response and almost twice this amount was spent on post disaster reconstruction and rehabilitation [4]. These numbers translate to false economy and cry out for disaster risk reduction funding. Developed countries, international organizations, and NGOs have invested little of what is needed in at risk countries for disaster preparedness, prevention, and recovery. At the least, this should include the implementation of good and enforced construction codes and environment protection legislation, in place warning and alert systems and protocols, and the elimination of corruption in many needy countries that works to defund disaster risk reduction guards.

# References

1. Murray, V., Aitsi-Selmi, A., & Blanchard, K. (2015). The role of public health within the United Nations Post-2015 framework for disaster risk reduction. *International Journal for Disaster Risk Science, 6*, 28–37.
2. Moon, S., et al. (2015). Will Ebola change the game? Ten essential reforms before the next pandemic. *The Report of the Harvard Global Health Institute and the London School of Hygiene and Tropical Medicine Panel on the Global Response to Ebola* (18 pp.).
3. WPRO (2015). World Health Organization Fact Sheets. http://www.wpro.who.int/mediacentre/factsheets.
4. Kellett, J., & Caravani, A. (2013). *Financing disaster risk reduction. A 20 year story of international aid* (50 pp.). Washington, DC: The Overseas Development Institute and the Global Facility for Disaster Reduction and Recovery.

# Chapter 4
# Conditions that Aggravate the Effects of a Hazard Event on Citizens and Property

## 4.1 Location

Location is an important factor that affects the impact of a hazard on society and its environment be it urban or rural. An impact is great when people are close to the point where the force of a hazard is strongest such as close to the epicenter of an earthquake, proximate to an erupting volcano, or in hilly terrain with a history of landslides. Citizens living in the path of a hazards such as a lava flow, a mud/debris flow, a high energy hurricane (typhoon, monsoon) or cyclone, or a tsunami are similarly in danger from the event. Population downwind of airborne industrial pollution or volcanic ash fall, or down flow of flood waters or contaminated effluents or groundwater can likewise suffer great harm. The threats to people can be magnified when the ecosystems around them are disrupted by hazards, albeit temporarily. Topographically, citizens living in urban centers, sometimes in valleys other times in areas of moderate to low relief, and encircled to some degree by industrial operations and with large volume vehicular traffic risk breathing harmful airborne particulates (<2.5 μm) and metals/chemicals aerosols. Depending on weather conditions, deadly urban smog can form that can sicken and kill people. Location of a population at or close to the outbreak (node) of an infectious disease puts people at risk of contracting the disease from airborne or contact transmission.

## 4.2 Population Density

The potential for the hazards to injure and kill people, damage or destroy property and infrastructure increases significantly in densely populated urban centers. This is the result of natural population growth from within, demographic changes as people move from rural settings to urban centers (in-migration, especially in Africa and Asia), immigration, and changes that are made to absorb an additional population.

© Springer International Publishing Switzerland 2016
F.R. Siegel, *Mitigation of Dangers from Natural and Anthropogenic Hazards*,
SpringerBriefs in Environmental Science, DOI 10.1007/978-3-319-38875-5_4

According to demographic projections previously cited, the 53 % of the world population that live in urban centers in 2015 (3.7 billion vs. 3.6 billion in rural settings) will grow to 70 % by 2050 (6.9 billion vs. 2.9 billion in rural settings) generally greatly increasing population density [1]. To accommodate growing populations, urban centers build within and upward where infrastructure can handle the additional citizen utility demands (water, gas, electricity, sanitation). They also grow outward and encroach on adjacent ecosystems with housing, employment edifices, and infrastructure expansion. The increases in population density makes more people susceptible to danger from any hazard event that strikes. As described in the previous paragraph, the dangers are related, to the geographic location of a population center. On the basis of historical hazard events, it is evident that geographic location influences the potential of disaster more so for urban centers and their growing populations as well as for citizens inhabiting rural agricultural zones.

# Reference

1. Demographia. (2015). *Demographia world urban atlas*, 11th annual edition, 2015: 1 (134 pp.). http://www.demographia.com/db-worlsua-pdf.

# Chapter 5
# Early Warning Systems (EWSs)

## 5.1 Alert to an Impending Disaster

Prediction of natural and human-assisted or intensified hazards is the basis for EWSs. We will give a general assessment of EWSs here and discuss their status in some detail later in the text when treating specific hazards. At present, for natural hazards, there are EWSs that can be put in place for flood hazards that are accurate in terms of where and when they will occur, the level above stream/river banks flood waters will rise, how long before a location will be flooded and how long flood stage will last. EWSs for volcanoes that are coming into an eruption stage can be established that are generally accurate as to when there will be an eruption, whether a volcano will launch tephra or emit lava and where it will flow, but can not accurately predict the force of an eruption nor how long it will last. There are EWSs signs of landslides that are related to topography, observations of slow earth movement, and amount of rainfall an area has received but these are often missed.

## 5.2 Status of Selected EWSs

### 5.2.1 No EWS for Earthquakes

There is no EWS for earthquakes. Precursors are being researched worldwide. A survey of 504 households in Teheran found that average households were willing to pay 367,471 Rials monthly (~U$S38 in 2007) for a hypothetical Earthquake Early Warning System (EEWS) [1]. Those more likely to pay were more educated with more children, and that possessed a home fire alarm. Studies like this one in other earthquake prone locations can help determine investment possibilities to mitigate the hazard potential for death and injury if an EEWS can be found. Although EEWS are not yet available, an EWS is in place in the oceans for an earthquake triggered

© Springer International Publishing Switzerland 2016
F.R. Siegel, *Mitigation of Dangers from Natural and Anthropogenic Hazards*,
SpringerBriefs in Environmental Science, DOI 10.1007/978-3-319-38875-5_5

hazard, the tsunami. When functioning as planned it can alert coastal residents to flee to higher ground or inland if the topography inshore is rather low lying.

The most recent effort to develop an EEWS proposes to use smartphones [2]. Researchers developed an algorithm (the APP MyShake) that uses smartphones as a global network that instantly report accelerometer signals and GPS locations from thousands of Android phones to give magnitude and epicenter of an earthquake of a magnitude of at least 5, 10 km away. The success rate they claim is 93 %. The MyShake App can be downloaded now. The EEWS could give people seconds before seismic waves arrive at a location. The researchers estimate that the early warning signal will give people 20 seconds to seek safety from a 7.8 magnitude event. Research will continue and is promising especially for countries that do not have extensive seismic detection equipment but where millions of smartphones are in use and if the early warning time can be lengthened and charged cellphones that are off can be activated to sound the warning.

## 5.2.2   Extreme Weather EWS

Meteorologists provide the EWSs for extreme weather conditions such as torrential rains that can give rise to flooding, high energy storms (hurricanes, typhoons, monsoons) and storm surges, and tornados (cyclones), smog, and heat waves. The alerts come via radio/television, telephoned alerts, or from police, firemen, or other municipality personnel. Citizens can prepare themselves and/or their property for the extreme weather conditions or evacuate away from them.

## 5.2.3   Infectious Disease EWS

An EWS for the onset of a node of an infectious disease requires that health officials communicate this outbreak to the WHO. The WHO then is charged with immediately notifying national health ministers worldwide about the happening. If the disease can be controlled from spreading by vaccines or medicines, these can be supplied immediately from WHO staged supplies or from stores of other nations if they are lacking at the outbreak location. Protocols are put in place for treating infected individuals and protecting the general public, and controls are set up at transportation sites or immigration sites to assure that travelers are not carriers of the disease. Without this early warning system a disease can spread as was the case with the SARS epidemic that began in China and when the government was late in reporting the disease to WHO. The MERS epidemic in Saudi Arabia was initially under reported until the disease was detected in people traveling from the infected area to other countries in the Middle East and elsewhere.

### 5.2.4   Pollution EWS May Be Problematic

There is no EWS for pollution that sickens populations and damages ecosystems and the organisms they sustain. The health threat becomes apparent, for example, when children and others in a population living close to or downwind of a battery factory or smelter get sick from heavy metal poisoning. The damage to ecosystems becomes apparent when there are fish kills from unprecedented algal growth stimulated by nutrient runoff into water bodies (e.g., streams/rivers, oceans) that upon death uses all water body oxygen during decomposition, or a release of neurotoxins from dead zooxanthellae, both of which kill fish. The damage becomes apparent when productivity from agricultural fields, lakes, and rivers is lost or greatly diminished from inflow of acid rain generated by coal-fired power plants that released sulfur dioxide into the atmosphere that reacted there to form sulfuric acid, the principal component of acid rain. Locally acid mine drainage can do the same. There is no alert to sudden industrial accidents that sicken or kill citizens or temporally destroy their livelihood and disrupt ecosystems such as the release of radioactive gases and particles at Chernobyl, USSR or Fukushima, Japan, oil spills in Alaska and the U.S. Gulf Coast, toxic chemicals release at Bhopal, India and release of toxic liquid mine wastes from storage ponds that are breached. We learn from events such as these why failure happened that impacted populations and ecosystems. Our efforts should be to try to institute technology (e.g., emission capture and control systems, scintillometers that detect and automatically sound siren alarms in populated areas when radioactivity is released). This alerts monitors that immediately warn populations of an imminent danger that will allow citizens time to rapidly evacuate what are expected to be danger zones. Governments can pass legislation and enforce laws that can protect ecosystems from long lasting disruptions.

### 5.2.5   EWSs Problems

There have been failures in the process of hazard monitoring, risk assessment of a hazard threat, forecasting tools used in the monitoring, and a system to put out a warning alert to the public as soon as it is issued. These are often caused by a lack of linkage and ready communication between the components of an EWS [3]. This emphasizes the need to and benefits of standardizing, and installing, and maintaining EWSs both nationally and internationally.

# References

1. Asgary, A., Levy, J. K., & Mehregan, N. (2007). Estimating willingness to pay for hypothetical earthquake early warning systems. *Environmental Hazards, 7*, 312–320.
2. Kong, Q., Allen, R. M., Schreier, L., & Kwon, Y.-W. (2016). MyShake: a smartphone seismic network for earthquake early warning and beyond. *Science Advances, 2*, e1501055. 8 pp.
3. Garcia, C., & Fearney, C. J. (2012). Evaluating critical links in early warning systems for natural hazards. *Journal of Environmental Hazards, 11*, 123–137.

# Chapter 6
# Damping the Dangers from Tectonics-Driven (Natural) Hazards: Earthquakes and Volcanoes

## 6.1 Tectonic Drive

Our habitat Earth has a radius of 6370 km (3960 mi). If we were able to see a section cut through the center of the Earth we would see that it is comprised of nearly concentric layers of materials with different measurable properties that allow us to divide it into several parts. The outer rocky layer is the crust that is from 10 to 40+ km (6–24+ mi) thick depending on whether you measure it from the ocean floor or under mountainous region. It is made up of several large bordered masses called plates. The crust is underlain by a mainly rock layer called the mantle that reaches a depth to 2900 km (1800 mi). The upper 100–250 km (60–150 mi) of the mantle is comprised of non-rigid material (perhaps molten-like in nature) at 1300 °C that underlies the crust and is called the asthenosphere. Beneath the mantle lies the iron-rich molten outer core that reaches to 5000 km (3100 mi) and the solid inner core that extends to 6370 km (3960 mi).

Tectonics is the study of large scale movement in the Earth with the rupture and/or deformation of the earth's crust. The frictional stresses that cause this are related to the movements of juxtaposed rigid sections of the crustal plates dragged along the non-rigid asthenosphere that caps the underlying upper mantle. The movements are likely caused by convection cells in the asthenosphere driven by heat flow from the underlying mantle. These follow a convection path as they rise, cool towards the crust and move laterally exerting a drag on the base of a tectonic plate, moving it a few centimeters or less annually. The cooling mass in the convection cell then moves down to reheat and continue the process. This is similar to what you would see looking into the side of a pyrex pot of boiling water. The plates have basic movements. One is drag in opposite directions of adjacent cells that pulls the plates apart and causes a rift as crustal rocks rupture. Another is push of one plate against another causing them to collide. In this scenario one plate can dive under the other (called subduction) or can override the other (obduction) or can push against the other causing it to yield and move up to form mountains. Still another basic

© Springer International Publishing Switzerland 2016
F.R. Siegel, *Mitigation of Dangers from Natural and Anthropogenic Hazards*,
SpringerBriefs in Environmental Science, DOI 10.1007/978-3-319-38875-5_6

movement is when adjacent plates slide laterally along their contact. The physical expressions of tectonics at the temperatures and pressure stresses exerted from plate movement are faults where there is a fracture and displacement of one part of a rock mass from another, and folding of rocks seen as convex and concave structures (anticlines and synclines, respectively). Activity at plate boundaries gives rise to earthquakes and volcanoes.

# Chapter 7
# Earthquakes

## 7.1 Prediction, Prevention

Earthquakes occur worldwide but generally recur and are most destructive at or near zones where plate tectonics are most active such as the Circum-Pacific "ring of fire" from Western South America north to Mexico, and Western United States to Alaska and then west to the Aleutian islands and south to Japan, Indonesia, and Southeast Asia and then past New Zealand and Antarctica back to Western South America. The Himalayan Asian belt that affects China, the Indian subcontinent and the Middle East represents the other major earthquake zone. Earthquakes damage and destroy property and infrastructure. They do not injure and kill people. Collapsing building, falling heavy ceiling material (often ceramic tiles), ejected concrete, brick, and rock facing, pieces of metal, and glass shards injure and kill people. Secondary events triggered by earthquakes do the same such as via fires (e.g., in Japan, San Francisco), disease (e.g., in Haiti, cholera), landslides and rockfalls (e.g., in Nepal, India, Pakistan), and tsunamis (e.g., in Indonesia, Japan). In the paragraphs that follow we will review what is being researched and what can be done to ease the burden of earthquake events on populations exposed to this natural hazard.

At this time there is no way to tell specifically where an earthquake will occur, when an event will strike (e.g., within a week, a year, a decade, or longer), and with what force (on the Richter scale) it will impact a location, nor the major type of movement, whether shaking, jarring, or rolling, or a combination of these. In 2008, on the basis of much research and an estimated 99 % statistical probability, the United States Geological Survey predicted that California would suffer at least one earthquake with a magnitude 6.7 or higher sometimes in the following 30 years. In 2015, the Geological Survey predicted that California had an 85 % statistical probability of experiencing an earthquake of magnitude 5 or higher during the next 3 years. The statistical analyses are based in great part on the recurrence intervals (how often an event of a given magnitude has occurred) of earthquakes in the past

F.R. Siegel, *Mitigation of Dangers from Natural and Anthropogenic Hazards*, SpringerBriefs in Environmental Science, DOI 10.1007/978-3-319-38875-5_7 and their magnitudes as interpreted from seismic records or in rocks that reveal past events (faults) that can be dated by geological research teams. This is an indication of the present status of earthquake prediction.

There is no way to prevent an earthquake. One concept that has been researched is based on the observations that water or other fluids injected into the subsurface during petroleum recovery or fracking for natural gas in the United States resulted resulted in low magnitude seismic activity. The question is whether injection of fluids to facilitate small slippages and minor earthquakes along an active fault such as the San Andreas fault in California can lessen the possibility of a high magnitude earthquake. The problem is to determine what density of fluid to use, how much to inject, and what pressure to use for fluid injection in order to cause low magnitude earthquakes to release stress at the fault without exceeding a tipping point and triggering a major earthquake.

## 7.2   Precursors for Earthquakes Being Researched

Nonetheless, research is on going worldwide on methods based on observations and measurements that by themselves or in consort may avail populations a degree of prediction of the possible onslaught of a damaging earthquake. To the present, although some of the prediction methods have had spotty success, none has been reliable in terms of foretelling when, where, and with what intensity an earthquake will strike. If one or a combination of precursors becomes reliable in the future, people may have time to gather important papers and family items together before evacuation if it is indicated. These are discussed in the paragraphs that follow.

### 7.2.1   The Gap Theory

The gap theory aims to predict where along an earthquake prone fault zone an earthquake may occur but not when nor with what force. It is based on the observations that where two plates meet to slip past one another they lock and cause an earthquake when stress breaks the lock and there a sudden movement and release of energy at what is called the epicenter. This may occur in rare cases of damaging earthquakes at recurrence intervals along a fault or plate boundary breaking locked sections and releasing seismic energy. When the locations of the epicenters of past earthquakes are plotted on a map along the slippage plane, there may be gaps. It is at a gap, a section that has not experienced an earthquake, that some seismologists expect the next movement to occur. If the rock rupture and release of seismic energy follows a recurrence interval, perhaps of 30 or 50 years or longer, for example, scientists can 'guesstimate' when the next earthquake may occur. If coupled with a gap situation, the when can be complemented with the where it is most likely to occur.

If there is a relative consistency of earthquake magnitude and motion in a seismic zone, a guess can be made of the expected force. This allows preparedness for the possible event (e.g., retrofitting structures).

## 7.2.2 Seismic Activity

Seismic activity—foreshock activity seems to be signaled by a decrease in seismic velocity ratio of P (pressure) vs S (shear) waves, Vp/Vs, for days before an earthquake and then change to normality just before a quake. P waves travel at higher velocities than S waves. Monitoring seismic activity records can alert scientists to the possibility of an earthquake but without establishing the when, where, and magnitude parameters with real time accuracy.

## 7.2.3 Change in Water Levels

Change in water levels in wells and nearby water bodies has been observed prior to an earthquake. Sometimes water levels fall, sometimes the water levels rise, and sometimes no change is observed. This is another possible precursor when coupled with precursor signals from other observations and/or measurements.

## 7.2.4 Anomalous Animal Activity

Anomalous animal activity during winter cold as rats, snakes, centipedes, and weasels leave their underground habitats. This was judged to be a precursor to an earthquake described in the following paragraph but was not observed as a precursor to other significant earth-shaking events. Research on the topic is mainly in China and Japan, countries where anomalous animal activity has been observed in some instances.

### 7.2.4.1 Prediction Successful, Prediction Unsuccessful

As noted above, prediction of a coming earthquake is not possible with any certainty as is a guess as to the probability of where, when, and with what magnitude one may occur. Nonetheless, there was a successful prediction in China in 1975 that saved hundreds of thousands of lives and at least as many serious injuries. During the winter, geologists and other scientist followed changes over a few months in the precursor activity cited above with relation to the city of Haicheng with its population of one million people: regional changes in seismic foreshock activity, changes in elevation of a nearby sea and groundwater levels, and anomalous animal activity that was reported by many citizens. On this basis, the government ordered the evacuation of the city. The day after the evacuation, an earthquake with a magnitude of 7.3 struck the city at about 7:30 in the evening. Because of the evacuation only 2000 people died and about 27,000 were injured. Without an evacuation, the Chinese government estimated that more than 150,000 would have been killed and injured. However, there were no precursors to the July, 1976 7.6 magnitude earthquake that struck the Chinese city Tangshan with one million people where 240,000 people died and 164,000 were injured. There may have been a factor related to the Haicheng earthquake occurring during the winter and the Tangshan earthquake during the summer but this was not evident to researchers except perhaps for the observed anomalous animal activity.

### 7.2.5  Strain Gauges

Strain gauges—strain measurements are made across a fault and a norm in the degree of stress is established. A change to a steadily increasing stress may be a precursor to an earthquake. This is likely the result of a "stretching" of the rock under increasing stress from lateral pressure that may either ease off or cause a sudden breakage at the fault: an earthquake.

### 7.2.6  Radon Gas Concentration Change

Radon gas concentration change—increasing numbers of small fractures that develop as stress along a fault zone increases provide pathways for enhanced escape of the radioactive gas radon. If an increase in the concentration of radon emissions can be isolated from other factors that can affect the emissions (e.g., increase temperature, atmospheric pressure), the measurement may be a useful precursor with others being evaluated to warn people of an impending earthquake.

## 7.3  Mitigation

### 7.3.1  The Earthquake Itself

Mitigation of the effects of the shaking, jarring, and rolling movements of a high magnitude earthquake (e.g., >6 on the Richter scale) to save people from injury and death, to minimize property damage or loss and infrastructure damage or destruction can be accomplished in several ways. First, new construction in known earthquake prone zones must follow building codes that use the best available techniques and materials that prevent building collapse (e.g., structural elements such as columns, beams, and flooring) and the loosening and ejection of facing and building ornamentation. This can meliorate injury and death because of collapsing buildings and ejected materials (e.g., facing, concrete, glass) that kill people during a strong motion event. The California and Japanese Building Codes, respectively, are excellent for setting and keeping up to date norms on construction in earthquake prone zones [1, 2].

For high rise structures, for example, structural skeletons must have extra steel bracing that is tied together using the best available bolting and/or welding techniques. An up to date building code also includes evaluating the foundation materials on which structures are built and the use of base isolation techniques (basically rubber/metal pad shock absorbers) on which the building is set so that the isolation units (energy dissipation methods) move with the earthquake motion reducing (damping) vertical and horizontal vibrations in the structure. Similarly, computer controlled weights on roofs of skyscrapers (Active Mass Driver) are used in Japan to dampen the motions produced by an earthquake. A damping method used in Japan for more than 20 years is being used for the first time in San Francisco, California, USA in the construction of a new multistory medical center. The skeleton of the building has 7 foot (~2.2 m) wide exterior wall panels with steel dividers (buckling-resistant braced frames) in them that are filled with a thick viscous synthetic rubber (polyisobutylene). The thick, viscous wall dampers were tested 6 years at the seismic testing facility at the University of California San Diego before construction. From the results of the tests, structural engineers calculated that the walls would absorb 80–90 % of an earthquake's energy and minimize the violent shaking, jarring, and rolling generated during an event, thus increasing the damping resistance of the hospital. The hospital would withstand a 7.8 magnitude earthquake, the same magnitude of the earthquake that destroyed San Francisco in 1906. To further reduce the impact of an earthquake, building codes must require automatic shut off of utilities' gas flow and electricity to reduce the possibility of fire and explosions that could be caused by their rupture during an earthquake. These and other adaptations of structural elements and retrofits of buildings are described in good detail in a U.S. Federal Emergency Management Agency handbook [3]. Japanese bullet trains and the BART trains in San Francisco have automatic stop sensors when there is a signal of seismic activity.

The retrofitting of existing structures is costly but will save lives and lessen injuries from violent earthquakes. It is in the interests of governments to help with grants, low cost loans, and tax incentives to encourage such an investment by owners. During October 2015, the Los Angeles city council enacted a seismic law that requires that brittle concrete buildings and wood-framed apartments built on top of carports be refitted to better withstand violent earthquake shaking, jarring, and/or rolling motions. During earthquakes in 1971 and 1994, these types of structures collapsed and killed citizens. The law requires that 13,500 wood apartments be refitted in 7 years and more than 1000 concrete structures be refitted in 25 years. The costs are estimated to be $60,000 to $130,000 for the wood structures and millions of dollars for the high concrete structures. The city council has yet to determine how these costs can be shared such as by relief on property and state income taxes or by an increase in monthly rents.

In addition to applying these mega-methods to minimize injury and death and loss of property, there are other relatively simple things that homeowners and businesses can do to minimize impacts from earthquake motions. These non-structural mitigation methods would include anchoring appliances and other heavy moveable items to a floor or wall (e.g., refrigerators, stoves, washing machines, dryers, dishwashers, or heavy furniture or other heavy free standing equipment) to prevent or limit their movement in a home or business. Bookcases and file cabinets can also be fixed in the same way. Kitchen cabinets and drawers should have firm latches to prevent dishes and other kitchenware from lurching out and crashing onto a floor or a person. These and other methods are described in excellent detail in a U.S. Federal Emergency Management Agency publication [4]. An investment of hundreds of dollars for these in homes and a few thousands for businesses can protect people from injury and save much money in material and property losses. Along these lines, in preparation for earthquakes in the future, citizens put mitigation actions (e.g., retrofit structures to latest building codes) in second place. They follow survival actions (e.g., have water, food, radio, first aid kit, flashlight ready) so that survival and contents security overrides cost considerations of retrofitting [5].

## 7.3.2   Earthquake Triggered Hazards

### 7.3.2.1   Tsunamis

Mitigation for earthquake triggered hazards to limit injuries and death and damage to property and infrastructure is possible for some of the hazards but not for others. This depends on the level of economic investment governments or international organizations are willing to support. For example, tsunamis are triggered by undersea earthquakes. Earthquakes are unpredictable so tsunamis are as well. In the same vein, tsunamis are not preventable, but an in place tsunami warning deep-sea detection network of systems of sea buoys anchored to the sea floor, can detect the onset of a tsunami using seismic sensors. The detection networks can establish where an earthquake occurs using GPS coordinates, can follow changes in the height of seawater above the seafloor, and direction and rate at which water is moving using pressure sensors and directional/flow rate sensors. These data are sent from the network of detection buoys to tsunami warning centers where they are processed with high speed computers together with additional data on seafloor topography (bathymetry) along a tsunami flow path. Warnings are sent instantly to coastal zones that predict how soon a sea wave moving at speeds up to or more than 650 km/h [over 400 mi/h] in the deep ocean will slam into a coast together with an estimate of the maximum height a tsunami wave will reach. The computer analysis will also determine when secondary waves will return and their expected heights. The entire process takes about 5 min. Alerts can be issued to coastal area inhabitants giving them time to evacuate to safe higher ground. Tsunami impact mitigation relies primarily on the warning methods used by police, firefighters, and civil defense personnel that gives coastal residents time to evacuate inland perhaps to 2 km, the maximum recorded reach inland of a mega-tsunami, depending on topography. Certainly, critical utility systems have to be protected in well anchored waterproof surface or underground enclosures that cannot be ripped up by a tsunami or short-circuited by seawater contact. For an event in not too distant deep sea, the detection-warning system may not be effective because of a limited time to receive an alert and evacuate. A detection network was not in place when the 2004 tsunami killed more than 230,000 people in Sumatra, Indonesia, and several other countries it impacted. A network for the Southwest Pacific Ocean is now in place. To some degree, a seawall high enough to damp the force of a tsunami can mitigate the damage it will do along a coast. An estimate of a height that can damp the force of a tsunami comes from geological studies of historical (or more recent) tsunami sediment deposits. At the Fukushima/Daiichi earthquake/tsunami/radioactivity release disaster in 2011, the "protective" sea wall had a height of 5.72 m in an area where an 869 AD tsunami reached a height of about 8 m with an estimated recurrence interval of 1000 years had been reported on in 2001 [6]. There is no doubt that if the Tokyo Electric Power Company engineers, consultants, and other policy makers had read and been aware of the article in the Journal of Natural Disaster Science [6] and built up the "protective" sea wall to 8 m or higher, the force of the tsunami would have been damped to a good degree in parts of the impacted coast. This would have caused less damage and destruction than was inflicted on the area by the tsunami, the release of radioactivity as gas and particles that poisoned an extensive zone away from the ruptured nuclear facility, and perhaps prevented the release of radioactive waters into the Pacific Ocean that is a continuing saga in 2015. The height of a sea wall that may be built will vary from one coastal location to another depending on the coastal topography and the hydrography of the seafloor.

### 7.3.2.2 Landslides

There is little that can be done to prevent landslides caused by the shaking, jarring, rolling motions generated by an earthquake. Analysis of world earthquake and land-slide data on facilities showed statistically that earthquakes that trigger landslides (cascade effect) kill more people than earthquakes alone [7]. Clearly, a hilly topography is susceptible to landslides. However, the susceptibility is controlled to a great degree by the earth materials comprising the terrain (e.g., soil, sediment, and rock type). Another influencing factor is the internal structure of the earth materials whether dipping towards a slope face (likely to slide) or into a slope. If an earthquake ruptures water pipes in a residential zone, the release of water into the terrain abets landslides because of the weight added to a slope by in seeping water, added water pressure, and the fluid lubrication imparted to the soil and rocks comprising the terrain. Mitigation to avoid landslide problems is not to locate growing populations where history, topography, weather, and geological characteristics of an area suggest future earthquake tremors or shocks will cause landslides. For existing populations in areas at risk for landslides, the vulnerability can be lessened by directing drainage away from a slope, by installing retaining walls with outlets for rain that might seep into the ground, and by the installation of buttresses of cylinders filled with concrete towards the base of a slope. In some cases, municipalities have used shotcrete to seal an area of hilly terrain underlain by earth materials susceptible to landslides so that precipitation could not seep in and destabilize slopes. This prevents landslides and protects populations from injury and death, and critical infrastructure from damage or destruction. It should be noted that some apparently solid sedimentary rocks become fluid like (thixotropic) when shaken by an earthquake. This leads to ground strength failure as well as destroying infrastructure that may have been inadvertently built on them such as the collapse of the ramp of a bridge from Oakland, California to San Francisco during a major earthquake in 1989. As with other disasters, if a population suffers being overrun by a landslide, preparedness with search and rescue teams, earthmoving equipment, and other needs described in the preparedness section can limit the deaths and injury from a landslide. Rockfalls can be a result of earthquake motions. The possibility of a rockfall can be lessened by keeping rock faces in place either with rock bolts or chain link retaining barriers.

### 7.3.2.3   Fires

Earthquakes can be responsible for fires when cooking stoves are overturned and ignite a home and when gas delivery pipes are ruptured and gas ignited by torn up electrical lines. Ruptured water lines interrupt the flow of water for firefighters. Mitigation of conditions that can ignite major fires from earthquakes include automatic shut off of gas flow when pipes are ruptured. This lessens the fire potential if electrical wire are live and exposed. A water supply independent of "normal" water lines is necessary to fight existing fires.

### 7.3.2.4   Infectious Diseases

Earthquakes can create situations where infectious disease can develop and rise to an epidemic because waste collection, water treatment plants, and sanitation systems have broken down. The lack of access to clean water and sanitation sets the stage for intestinal illnesses (dysentery) and infectious diseases (e.g., cholera) ravaging a population as was the case for years subsequent to the 2010 Haitian earthquake. A response to the lack of safe water and the onset and spread of disease is to initially supply people with tablets that purify otherwise contaminated water and/or distribute of bottled water and the rapid set up of sanitation stations for the public where wastes can be disinfected. These actions can mitigate health threats for a community until water treatment plants and sanitation systems are back in operation. If some people become infected with cholera, they should be quarantined and treated thus stemming possibilities of transmission to the general population. Dead bodies that are the result of earthquakes are sources of disease and have to be interred as soon as possible.

## 7.4   Preparedness

The preparedness for an earthquake is much the same for other major disasters. The preparedness process is done within the context of not having reliable predictions of when an earthquake will occur whether in years or months, without knowing the magnitude of the event nor its epicenter and reach, nor the probability of a failed prediction. Nonetheless, preparation in a known earthquake prone zone is an absolute necessity to lessen the earthquake impact on life and property. All of the preparedness programs require funding whether from a nation itself or from financially strong nations to low income, less developed ones. Certainly, the retrofit of structures and especially infrastructure that is not built to an up-to-date building code should be a major phase of the preparedness process [8]. Unfortunately this is not often done because of lack of economic resources. Economically viable and a life-saver is a continuing public education and awareness program so that people know how to respond when they feel the earthquake motion. It is necessary to have planned for emergency medical care (hospitals, clinics, field hospitals) with ready access to medicines and vaccines, food, clean water, access to adequate sanitation, shelter (tents), clothing, communication equipment, transportation as needed, and fire-fighting equipment. Earthquake preparedness differs to some degree by the need to have access to equipment to lift and move collapsed building sections and search and rescue teams with canines trained to search for trapped, perhaps unconscious citizens. There should be teams with the capability to reach geographically inaccessible areas (helicopter support) as was the case following the Kashmir earthquake in 2005 that killed 30,000 people and that destroyed transportation routes to isolated and remote Himalayan villages that needed help. This scenario repeated itself on April 25, 2015 in Nepal when a 7.8 magnitude earthquake caused 9000 deaths and 23,000 injuries in Katmandu and surrounding areas from collapsing buildings especially of older construction, and triggered landslides, rockfalls, and snow avalanches with isolation of Himalayan villages. International support flowed into the area with medical care, food, clean water, tents and blankets for shelter, specialized personnel for search and rescue, and helicopter support to bring aid villages where road access was not possible.

## References

1. California Building Standards Commission. (2013). *California existing building code* (752 pp.), Vols. 1 and 2. Sacramento, CA: California Building Standards Commission.
2. Teshigawara, M. (2012). Appendix A: outline of earthquake provisions in the Japanese Building Codes. In *Preliminary reconnaissance report of the 2011 Tohoku-Chica Taiheigo-Oko earthquake, geotechnical, geological, and earthquake engineering 23* (pp. 421–446). Tokyo: Springer.
3. Federal Emergency Management Agency (FEMA). (2002). *Earthquake hazard mitigation handbook* (Au.: Structural elements in buildings). Washington, DC: unpaginated.

4. Federal Emergency Management Agency (FEMA). (2005). *Earthquake hazard mitigation for nonstructural elements* (50 pp.). Washington, DC: FEMA.
5. McClure, J., Spittal, M., Fischer, R., & Charleson, A. (2015). Why do people take fewer mitigation actions than survival actions? Other factors outweigh costs. *Natural Hazards Review, 16*, 04014018.
6. Minoura, K., Imamura, F., Sugawara, D., Kono, Y., & Iwashita, T. (2001). The 869 Jogan tsunami deposit and recurrence interval of large-scale tsunamis on the Pacific coast of Northeast Japan. *Journal Natural Disaster Science, 23*, 83–88.
7. Budimir, M. E. A., Atkinson, P. M., & Lewis, H. G. (2014). Earthquake-and-landslide events are associated with more fatalities than earthquakes alone. *Natural Hazards, 22*, 895–914.
8. Davis, C., Keilis-Borok, V., Molchan, G., Shebalin, P., Lahr, P., & Plumb, C. (2010). Earthquake prediction and disaster preparedness: interactive analysis. *Natural Hazards Review, 11*, 173–184.

# Chapter 8
# Volcanoes

There are hundreds of active volcanoes on earth, most of them in the Circum-Pacific "ring-of-fire" at tectonic plate boundaries where a sea plate moves under a land plate at what is called a subduction zone. A volcano is considered active if it has erupted at least once in the past 10,000 years. Japan has 110 active volcanoes, Indonesia has 127 while Chile lays claim to 123, and Central America has 45. Volcanoes also occur where plates pull away from each other such as at the mid-Atlantic Ocean ridge or slide past one another (e.g., Hawaii). Geologists study volcanoes and the materials they deposit (ash, lava) in order to determine when a volcano erupted in the past and from these data how often. In many cases, where delicate monitoring equipment is in place, the onset of volcanic activity is often predictable.

## 8.1 Prediction, Prevention

Obviously, when wisps of steam are seen emitting from a crater or from fissures on the flanks of a volcano, people are alerted to a phase of activity. Ground and adjacent water body temperatures are measured in situ *or* can be measured by infrared air photos taken from a small plane. If the temperature increases markedly above what is considered baseline (normal), this suggests that underground magma that could drive an eruption is moving towards the surface. Tilt-meters in place on a volcano's slopes can detect surface deformation caused by magma pushing up, another indicator of a possible eruption. Research is active on networks of GPS equipment to do the same. "Sniffers" or gas detectors at a crater or on a small plane that overflies a crater can determine if there is a notable increase in release of gas such as sulfur dioxide above normal. This is another precursor to an impending eruption. The most important precursor to a possible eruption come from monitoring a volcano's seismic activity as magma moves in the subsurface causing cracks in the enclosing rocks. A marked increase in the low frequency seismic signals is a

© Springer International Publishing Switzerland 2016

F.R. Siegel, *Mitigation of Dangers from Natural and Anthropogenic Hazards*,
SpringerBriefs in Environmental Science, DOI 10.1007/978-3-319-38875-5_8

precursor that suggests that an eruption is imminent, probably within 24–72 h although there is no predictable exact time. The precursors give no indication of the force or volcanic energy that will be released nor how long an eruption will last. There are studies that suggest that anomalous animal activity signals a coming eruption but this is anecdotal and has not been proven scientifically.

Unfortunately, the equipment and personnel to monitor volcanic activity are not in place on hundreds of active volcanoes. Even with monitoring, an eruption may be so sudden so as not to present precursors. This was the case on Mt. Ontake, Japan, September, 2014. Monitors saw slight changes in seismic activity but not strong enough to warrant an alert. Hikers on the volcano were hit with a sudden ash flow, toxic gases, and a rain of rocks that killed 71 persons. There was a phreatic eruption when magma and superheated steam found a crack in the cap rock and burst through without warning rupturing the rock, and releasing the ash and associated gases.

## 8.2   Mitigation: The Volcano Itself

There is no way to prevent a volcanic eruption. There are ways, however, to mitigate the possible impacts from volcanic eruptions. These are important to the more than 500 million people that are at risk from active volcanoes. As with other primary or triggered disasters, mitigation to minimize or prevent injuries and deaths in at risk zones when there is an alert of an imminent threat of a volcano eruption is to evacuate citizens to safe locations where food, water, shelter, and medical attention are in place. Such preparation, when available, is a result of good government planning. In order to protect structures from glowing ash and tephra, building codes should require that a building's roof and outer shell be of nonflammable materials. In addition, where geological evidence shows that ash fall is dominant, buildings should have high pitched roof to obviate collapse from the weight of a buildup of ash especially if rain adds to the weight as might be the case on a flat roof top. Where topography and geological data suggest routes that lava flows are likely to take during an eruption, deflection barriers can be set in place to try to direct a flow away from villages or infrastructure. This is guided using hazard zone maps prepared by geologists and based on the field work they do on rocks that reveal a volcano's history of eruption activity.

## 8.3   Volcanic Triggered Hazards (Mudflows, Fires, Flooding)

Triggered events that result from volcanic eruptions can be more deadly to nearby and downwind populations than ash falls or lava flows. When masses of ash mix with water from rain or melting ice and snow at the upper slopes of a volcano, a mudflow may be generated that moves downslope at great speed following the topographic lows. The mudflow picks up sediment and rock along its route and can bury population centers in its path under meters of mud, killing thousands of people. Such was the case in 1985 in Columbia when a mudflow resulting from melting of snow and ice on the erupting Nevado del Ruiz volcano and the meltwaters mixing with underlying soil and rock released a mudflow that buried the city of Armero under meters of earth materials killing 23,000 of 29,000 residents. The volcano had been quiescent for 69 years. To try to avoid injury, death, and destruction from mud-flows means that human settlements should not be located where the topographic low areas are pathways for volcano triggered mudflows. A mega-mudflow (named Osceola) moved from Mount Rainier in Washington State 24 miles (38 km) towards Puget Sound more than 5000 years ago. Today there are many small cities located on the path of the mudflow (and others that preceded it) putting their populations at great risk if a mudflow of the same magnitude were generated again by activity on the volcano. It should be noted that mudflows that run into a river can dam it causing flooding upstream, and when the dam is breached, flooding downstream. This has to be planned for if the situation were to arise, perhaps by blowing up the temporary dam before it backs up river waters and causes upstream and downstream damaging, destructive flooding.

Forests can be leveled and set on fire by radiative blasts from a violent volcanic eruption that are directed laterally rather than upward. There is no way to mitigate such an action that would be destructive for an ecosystem. However, 10 years after the 1980 Mount St. Helens eruption in northwest United States, the forested areas that had been leveled by such a blast were rejuvenated with vegetation, animals, birds, insects, and other life forms that escaped the blast by leaving the area or burrowing underground and later reappeared and reestablished the natural ecosystem.

## 8.4 Preparedness

Preparedness lies first with a monitoring and an early warning system, second with planned evacuation routes, and third, as stressed previously for other disasters, with safe sites ready to receive and care for the evacuees. Geological scientists can study the recorded history of volcanoes and the history as revealed by ancient ash and lava deposits. From studies of the ages of eruptions, geologists can determine whether there is a recognizable recurrence interval and use this information as they monitor a volcano's activity patterns. For example, the Merapi volcano, Java, Indonesia erupts explosively every 5–10 years so that monitors know approximately when the next eruption is likely to occur and plan accordingly. Other volcanoes have no discernible recurrence history so that monitoring alone is the key to alert citizens to the probability of an eruption and the need to evacuate. The monitoring may be simple such as having volcano watchers that know a given volcano's signs of a pending eruption. This is used in many developing and less developed countries that are threatened by volcanic activity. Ideally, an electronic monitoring system and trained personnel to evaluate data on seismic activity, temperature changes, deformation signals, and gas concentration data is preferred but is costly to set up, operate, and maintain. In some cases, the benefits from the implementation of such a system have far exceeded its cost both in injuries and deaths and in material objects. For example, a system set up to monitor the Philippines Mount Pinatubo cost $56 million. Alert to the imminent eruption of the volcano in 1981 is estimated to have saved more than $500 million in property (e.g., airplanes) and possibly more than 5000 lives [1].

# Reference

1. Newhall, C., Hendley, H. J. W., & Stauffer, P. H. (1997). *Benefits of volcano monitoring far outweigh costs: the case of Mount Pinatubo*. U. S. Geological Survey Fact Sheet 115-97.

# Chapter 9
# Lessening the Impacts from Non-Tectonic (Natural) Hazards and Triggered Events

## 9.1 Floods

Floods are a global problem. They are predictable to some degree by weather fore-casting but to a greater degree and with more accuracy when drainage basin moni-toring equipment is in place. This includes stream gauges that telemeter the elevation of stream/river surface in a channel and the rate of water flow to a central computing station. The computed data from the telemetered sites plus the input of stream/river channel cross-sections data allow prediction of where flooding will be a problem, when the flooding will reach an area, and to what level out of a channel (magnitude) the flood is estimated to reach. This gives the populations at risk of the flooding early warnings (hours, days) and time to prepare for the floodwaters or to gather important documents and evacuate to safe higher ground.

Floods are not preventable but their effects can be mitigated to a good degree if flood control management is in place. Flood control is achieved in several ways so that waters can be retained and released slowly after the threat has passed (dams), or kept from moving out of their natural channels (levees, dikes). In addition, existing channels can be modified so that they can carry greater volumes of water without overflowing their banks and move more quickly through flood zones (deepen, widen, straighten channels), and by adding a flood wall to retain waters that exceed volumes carried in the modified channels. Sandbagging around structures may keep floodwaters from invading them.

Floods are often recurring events although their magnitudes may vary greatly. The recurrence intervals between floods of given magnitudes can be determined from an analysis of historical and modern records. These are generally reported in terms of probability of flood levels to be reached every 100 years, every 200 years, every 500 years (1 %, 0.5 %, and 0.2 % probability, respectively) and so on. These estimates are only suggestive but are used in determining whether building permits

© Springer International Publishing Switzerland 2016
F.R. Siegel, *Mitigation of Dangers from Natural and Anthropogenic Hazards*,
SpringerBriefs in Environmental Science, DOI 10.1007/978-3-319-38875-5_9

will be issued for residences or other structure for areas proximate to rivers, whether flood insurance for structures will be written for a given zone, and if written, what the cost of flood insurance will be.

Floods can cause landslides and potential health problems for populations when there is torrential rain that is long lasting. First, depending on the location of a flooding river channel, rushing water can undermine bank material or erode base of valley walls causing landslides that could affect people living in the threatened areas. Second, torrential rains can overload sewer systems so that sewerage carrying pathogens are discharged into waters that may affect water users downstream. Populations warned of these potential flood-triggered hazards can follow advisories to evacuate if necessary and drink bottled water if their water supply is tainted.

Preparedness against floods includes warning/alert systems sent by weather bureaus and flood management agencies and, of course, evacuation plans showing routes to safe sites with staffs and supplies in place to help displaced citizens.

An Intergovernmental Panel on Climate Change report suggests that flooding has been more frequent and more severe in some regions and that there is less flooding in other regions [1]. The report projects that extreme precipitation events will occur over most of the mid-latitude land masses and that those over wet tropical regions will become more intense and frequent. These extreme precipitation events often cause flooding that endangers people and property especially if early warning systems are not in place.

Natural changes in river flow other than those that result in flooding can be categorized as disasters for people and for a country. For example, the Semliki River marks the border between Uganda and the Democratic Republic of Congo (DRC). In recent years, and perhaps as a result of global warming/climate change, the volume of water discharged into the river from melting snow and ice from mountain peaks has been reduced as the ice has receded and less snow has fallen. In addition, there have been periods of erratic rainfall and the wet season is wetter. The result has been a meandering of the river that has flowed through Uganda fields and left them as part of the DRC. Ugandan farmers and herders have to pay DRC owners of the "new land" a tribute to be able to cross 15 m of river water to continue to work the fields and care for their herds on the other side of the river that was Ugandan territory before the border shifted as the river meandered. There is international intervention that is working to set a fixed border so that land that was lost can be assigned correct ownership and Ugandan territory can stop shrinking [2].

## 9.2    Mass Movements

### 9.2.1    Landslides

Landslides also called landslips are the best known of the disaster class called mass movements. Mass movements are a world wide problem that cause 4500 deaths and hundreds of millions of dollars damage each year to homes, infrastructure, and utilities. Other movements in this grouping include: (1) subsidence or a lowering of an area of the Earth's surface; (2) collapse of soil/rock into a void in the subsurface; (3) rockfalls; and (4) mudflows (debris flows). The latter is activated when torrential or sustained rain over a period of time loads water weight into slopes and seeps into earth materials there, lubricating them and also pushing grains apart because of increased water pressure. The heavy, lubricated, destabilized matter finally breaks loose and speeds downslope without warning at as much as a mile a minute (96 km/ph). The destabilization can be further abetted when some of the earth materials comprising a slope is composed of the clay mineral class called smectite (montmorillonite/bentonite), a mineral that expands when it is wetted and shrinks when it dries. A combination of the expansion and lubrication from added water and its weight can bring down a hillside. Populations at the base of slopes and beyond are at risk of sudden burial with major loss of life. It should be noted that away from hillsides, soils that expand when wet and shrink when dry cause cracks in basements and foundations of buildings that rest on them, a property damage problem but a problem that is not a threat to the well being of people. Avoidance of possible building sites underlain by expansive/shrinkage soils or extraction of the soils prior to construction can mitigate the potential basement/foundation problem. Keeping the soils wet with an irrigation system that activates during dry weather has been used to stabilize the reactive soils.

### 9.2.1.1  Observation and Measurements that Reveal Landslide Prone Regions

Landslides occur in hilly topography (elevated terrain, >15° slopes) where the soils are well developed, where poorly consolidated sedimentary rocks comprise the hills, where the structure of the rocks dip (angle) downward and outward from the at risk slopes, and where there is considerable rainfall. These are predictive observations that suggest the possibility of a landslide prone region. Landslides are a threat when water seeps into the hills adding water weight that increases the pull of gravity on the earth materials, increases groundwater (pore) pressure that pushes grains apart, and acts as a lubricant for subsurface earth materials. Other observations that alert citizens that there is a downslope movement of soil are cracks in the ground and trees that tilt back into a slope. A geologist's observations of scarps in a hilly area attests to it being landslide prone. Scarps are mini-cliffs often with exposed soil caused by a landslide but that may be hidden from the untrained eye by overgrowth. Tilt-meters (inclinometers) are electrical driven pieces of equipment that can be installed in slopes to telemeter movement or deformation data that suggests the possible onset of a landslide. A recent approach to identify the strong probability of a landslide and thus could serve as an early warning system that allows people to evacuate to safety uses fiber-optic strain sensors complemented by rainfall data [3]. These sensors render continuous monitoring in time and space. Fiber-optic sensors that can detect displacement, groundwater pore pressure, displacement (slow slope slippage), ground vibrations, and temperature, are glued equally spaced to the surface of a PVC (flexible) pipe and embedded in shallow trenches in a soil. The pipe can bend or twist with pre-slope failure tensile strain (e.g., elastic, plastic, shear, viscous volumetric) registered as 3D deformation. The various sensors will register the location of the deformation on monitors at an off slope location.

Physically, a landslide appears as a block of earth materials that breaks from a slope along a scarp and slides down along a concave surface pushing lobes of slide materials out at the base of the block. The areas affected by landslides are generally limited and distances landslide move are generally small although in areas of very steep topography the momentum of a landslide can carry the detached mass a significant distance and across adjacent terrain. As noted previously, landslide moving into a river valley can act as a dam, cause flooding initially upstream and subsequently downstream when the temporary dam is breached.

### 9.2.1.2   Human Activity that Can Cause Landslides

In addition to landslides starting as a response to an earthquake's shaking, jarring, rolling motions and excessive rainfall, human activity can lead to landslides as well. This can happen when toes of slopes that give stability to hills in landslide prone regions are cut away to establish a road. This disturbs a slope's equilibrium and is an example of bad land use planning that creates conditions that can lead to landslides. Similarly, landslides may be caused if the head of a slope at equilibrium between gravity stress trying to pull soils and rocks down and the strength of the mass resisting slope is weakened by overloading the head with housing and well-travelled roads. The added weight (stress) and vibrations from vehicular traffic can result in a slope failure. Also, sliding can be abetted by loss of anchoring vegetation especially where logging has been active.

### 9.2.1.3   Protection Against Landslides

In landslide prone areas as evaluated by geologists and geological engineers, citizens can be protected to a good degree from injury, loss of property, and even death in two ways. One is zoning that prevents habitation where conditions are conducive for landslips if a slope were to be loaded with destabilizing weight such as the earlier cited homes and well trafficked roads. Second, for existing habitation or planned habitation, geological engineering can diminish the threat of landslides. This is done by (1) redirecting rainwater or snowmelt so that it does note invade slopes at risk; (2) by installing retaining walls with perforations that allow in seeping water to exit a slope; and (3) by installing concrete caissons, a few feet in diameter down to base rock along the front of a vulnerable slope. These efforts are costly so that economic constraints prevent their general use. In areas of Japan where topography is hilly and where landslides threaten infrastructure including critical transportation systems, the Japanese government has invested in landslide prevention. They sealed hills with shotcrete so that rainwater or snowmelt can not seep into the hills and destabilize them with added weight and lubrication at earth material slide planes. Preparedness would require that earth moving equipment and search teams be available to rescue people who survived but are trapped by displaced earth materials.

Recurrence of landslides is associated with the amounts of rainfall an area receives over a period of time. Landslide activity will increase in some global areas and in some regions as a result of increased rainfall because of climate change. As mentioned in the previous section on flooding, extreme precipitation events, or simply increases in annual mean precipitation is likely for the high latitudes, the equatorial Pacific region, and mid-latitude wet regions and can provide the stimulus for landslide activity. Central America and South America are especially ar risk [1].

## 9.3  Avalanches: Mass Movements of Snow in Alpine Settings

An avalanche is a large mass of snow that suddenly slides downslope. There are two general types of avalanches: (1) sloughs that are small flows of powdery snow that are unlikely to kill people or destroy structures; (2) full depth avalanches that are massive slabs of snow that break loose from a mountain and cause death and destruction. The latter may carry ice, soil, and rock debris. Avalanches generally occur without warning on slopes >30° and <45°at alpine areas worldwide (e.g., Swiss mountains, Western Canada, New Zealand, Alaska, the Himalayas). One or a combination of factors can contribute to an avalanche event. These include storminess, slope shape, orientation of steep slopes with respect to the sun, the rough or smooth character of the ground beneath a snowpack, vegetation, the nature of the layers of a snowpack, and vibration. Thus, if there is a 30 cm (12 in.) or more snowfall in a 24 h periods, there is likely to be an overloading and an avalanche depending on how the layers in a snowpack are bound together. Most avalanches occur during snow storms and blizzards. Clearly, a steeper convex slope is more conducive to avalanche activity. A rapid temperature increase can cause melting of a layer in the snowpack that results in an avalanche. The more vegetation there is deters the down flow of an avalanche as would a rough rocky surface beneath a snowpack. Vibrations from activity on a slope (skiers, snowboarders, snowmobiles), thunder storms, low flying jet planes, and explosions can set off an avalanche. During WWI, thousands of soldiers in Alps regions were killed by avalanches triggered by artillery fire. Avalanches can flow downslope at speeds of 130 km/h (80 mph) or more. Depending on the mass being moved, an avalanche can kill people, and damage or destroy structures and infrastructure (homes, recreational areas, bridges, tunnels [block road and/or railway movement]), pipes and utility lines (water, natural gas, electricity), and put workmen maintaining an infrastructure at risk. Hundreds of people killed by avalanches each year. During 2014 and 2015, avalanches in Nepal, triggered by an earthquake and by unseasonal severe rain and snow blizzards, roared downslope from Mount Everest and other peaks in the region. Many trekkers were killed, as were Sherpa guides and climbers preparing for the climb to the summit of Mount Everest.

### 9.3.1 Protection from Avalanches

People can be protected by zoning that evaluates the history of avalanches, the paths they follow, and their frequency and reach in an alpine area to prepare risk maps to prevent use or allow limited or full use of the zoned terrain. In areas where use of terrain is allowed, warning systems are in place so that when sounded, citizens will not enter a threatened area, or if there, evacuate it immediately. Avalanches can not be prevented but defensive structures can be used in established populated areas to try to divert them such as snow fences or snow walls. Avalanche sheds can be used to protect structures and transportation routes by making the flow of snow ride over them. Dangerous buildups of snow on slopes that are known for avalanches can be set off as snow slides or snow slips using vibrations caused by controlled explosions with explosives implanted in the snowpack, dropped by helicopter, or delivered by artillery shells. Excellent sources of information on avalanches can be found at *www.ussartf.org/avalanches.htm* and at *www.conserve-energy-future. com/types-causes-effects-of-avalanches*.

## 9.4 Rockfalls

### 9.4.1 Conditions that Favor Rockfalls

A rockfall happens when a mass of rock from a very steep to vertical cliff detaches from the face of a cliff and free falls down. The rocks bounce off underlying rocks, often detaching them as well. The falling rocks crash onto the base of a cliff sometimes running out damaging structures and/or infrastructure, blocking roads and putting vehicles at risk. In addition to steep topography, the geological character of the rocks comprising a cliff, the climate, and sometimes vegetation are factors that influence the risk of a rockfall but do not predict when one may occur. Fractures and fissures in rocks can fill with water during a winter day and freeze at night causing ice wedging that weakens rocks against the pull of gravity. Trees that root in cracks and crevices in a rock cause root wedging that does the same. Stress when an earthquakes strikes an area can loosen rocks to the point that they fall from a cliff face.

### 9.4.2 Rockfall Prediction, Protection for Citizens

A geological evaluation of an area can reveal areas that have suffered rockfalls and areas likely to have rockfalls but geologists can not reliably predict when a fall will occur. To protect an area and its inhabitants, infrastructure, and businesses from rockfalls, municipalities have options. They can install catchment fences at the base of cliffs to prevent run outs, require that rock bolts be inserted to stabilize an at risk rock face or that chain-link fencing be fastened to the rock face.

## 9.5   Subsidence

### 9.5.1   Cause of Subsidence

Subsidence of an area of the Earth's surface is the result of the continuous extraction of large volumes of groundwater or petroleum from underlying sedimentary rocks without recharge or replenishment of fluids. Fluids in subsurface rocks provide buoyancy pressure that strengthens the resistance of the rocks to compaction and hence subsidence of the Earth's surface. If volume loss does take place in subsurface rocks, subsidence may or may not occur depending on the strength and thickness of overlying rocks. Overlying rocks may be inherently strong enough so that they do not subside. Conversely, they can respond to the compaction of underlying rock by subsiding.

### 9.5.2   Mitigating Subsidence

Subsidence can be arrested or even reversed somewhat in some geological settings if a degree of buoyancy pressure is reestablished by re-injecting fluid (e.g., brine) into the rock. For example, in oil fields, eight barrels of brine are extracted with one barrel of crude oil. The brine can be recycled into the oil reservoir under pressure. Similarly, if extraction of groundwater from aquifer rock is replenished by groundwater recharge, any subsidence that has taken place should stop. Oil production at the Wilmington Field in southern California, USA, began in 1938 and caused a subsidence that by 1958 reached 9.5 m (31 ft) at Long beach Harbor and extended to parts of Los Angeles Harbor. The subsidence damaged oil wells, pipelines harbor infrastructure, railroad tracks, streets, and bridges and reversed the flow of sewers and storm drains. Repair cost more than U$S 100 million at that time. Brine re-injection was used to arrest the subsidence and there was stabilization and a rebound of 30 cm (12 in.) [4].

Mitigation of a possible subsidence problem can be determined by geologists and geological engineers who study samples of the sequence of rocks from the surface to the oil or water reservoir. They can predict whether or not subsidence will take place and if it is realistic to plan to recharge the extraction reservoir with volumes of fluid that equal the volumes withdrawn, thus maintaining buoyancy pressure in the reservoir/aquifer. This can prevent or at least minimize subsidence if it begins.

### 9.5.3   Economic Problems from Subsidence

Governments can be stressed two ways economically because of subsidence. First is the cost to repair the damage done to infrastructure and structures in the areas affected by subsidence. Second are the economic losses that can be incurred if a productive sector is slowed down by not being able to extract critical fluids from the subsurface without having to invest more funding to improve extraction by drilling wells deeper. This would likely mean an increase in the price of a commodity used to make or grow a product and hence be inflationary for the public. A case in point is the more than four year severe drought being suffered in 2015 in the Central Valley (especially the San Joaquin Valley) of California, USA where subsidence has been a recurring problem during past droughts as groundwater was extracted, but not excessively, from aquifers to make up for the shortage of surface water. However, excessive groundwater pumping during the existing extended drought, mainly by the agricultural sector, has lowered groundwater tables to 100 ft (~30 m) lower than recorded in the past. As a result some areas experienced a subsidence of up to two inches (5 cm) a month as fine-grained layers in the aquifer were depleted of their buoyancy pressure and compacted. If the compaction is tight enough, part of the storage capacity (porosity) and permeability (ability to transmit fluids) of an aquifer could be lost. The subsidence varies with location in the valley. One area subsided at 1/2 in. a month. The maximum subsidence was about a foot (12 in. or 30 cm) a month [5]. Another result of the over pumping is a reduction of fresh water pressure in the aquifer that could result in salt water intrusion where the aquifer extends to the marine continental shelf.

The differential subsidence in the San Joaquin Valley has caused damage to infrastructure, the most important of which may be the California aqueduct comprised of canals, pipelines, tunnels and pumping stations. The aqueduct carries water from northern California rivers that receive melt from the Sierra Nevada snowpack and from rainwater some 400 miles (~600 km) to southern California. Change in the fall (inclination) of sections of the aqueduct from subsidence and low spots on the system prevent the water from flowing as well as it did pre-subsidence and thus needs reworking. Changes in the fall of sewer lines have to be repaired to prevent sewage backup and its consequences from a reversal of slope. Similarly, water pipelines have to be reset to allow efficient flow of water. Some bridges have subsided to the degree that they are no longer above the water surface. Roads are ruptured and have to be repaired. Levees in place for flood control when the rains come have sunk and have to be raised. Building foundations sink as well and need correction. Very important in this agriculturally dependent valley and its towns has been the destruction of thousands of public and private well casings. California and the agriculture industry have the economic wherewithal to make the necessary repairs once the drought breaks. The Central Valley grows about 50 % of the vegetables and fruits sold in the United States. As previously indicated, if crops fail or yields drop because of the lack of water for irrigation, the cost of the produce and fruit will increase. The state is investing large sums of money to develop a capital improvement program.

## 9.6  Collapse/Sinkholes

### 9.6.1  How Sinkholes Form

Collapse of a small areas of the Earth's surface into voids in the subsurface causes sinkholes. This is an action that takes place most often where there is a relatively high water table in terrain underlain by limestone, a rock type that houses many famous cave systems worldwide. In this scenario, groundwater moving slowly through limestone continually during geologic time (millions of years) dissolves the limestone leaving voids in the subsurface. When these are large enough and the roof rock strength cannot resist the pull of gravity, the roof rock collapses into the subsurface cavity. The areas affected are generally small, rounded or oval shaped, and tens of meters or less across. Sinkholes have caused structural damage to buildings, swallowed homes and car dealerships, ruptured roads (pavements) including highways, and in rare cases have caused injury and death. Sinkholes have developed as well in terrain underlain by salt-rich (evaporite) strata that are dissolved by irregular groundwater flow. In some cases, water pipes underlying roadways have broken and the released water has washed out the earth materials around a ruptured waterlines creating open spaces in the subsurface into which earth/road materials have collapsed. Underground coal fires can create cavities into which the overlying rock and soil can subside and ultimately collapse.

### 9.6.2  Minimizing Risk to Land Used for Housing and Development

To lessen the risk of suffering loss from collapse from a future sinkhole when buying a home or business or land to develop and build on, one should commission an evaluation of the terrain by experts that will predict its vulnerability for collapse. For example, a geologist will first assess the geology of the area under study (especially when underlain by limestone or dolostone), the topography, and determine the position of the water table that could cause dissolution of underlying rock. He/she will review the history of insurance claims against collapse situations in the area. In a surface analysis, the geologist will look for signs of potential surface movement that could portent sinkhole activity by examining building foundations in the neighborhood for visible cracks, especially arcuate ones, cracks in roads or sidewalk pavement, and depressions or low spots in the terrain as well as the presence of small ponds that could represent former sinkholes. This study will suggest the level of risk in an area but only an investigation of the subsurface conditions can give more accurate vulnerability information. This can be done using ground penetrating radar (GPR) that takes a short time using modern equipment that digitizes data for ready presentation and interpretation [6]. In one Florida study, the GPR data for nine 60 m lines were generated in 2 h. Another option is to do a seismic study that will yield information as to the solid rock nature or void presence in the subsurface. Studies such as these, where collapse is an endemic problem, but on a larger, perhaps county scale, can result in sinkhole vulnerability maps that can guide land use planning for small scale evaluations of potential sites for buildings or infrastructure. Where there is a void in the subsurface but there is a need to use the terrain, it may be possible to fill the void with grout to stabilize the surface but this can be very costly.

### 9.6.3   Prediction/Mitigation

A question exists as to whether scientists can spot an incipient sinkhole as it develops so as to be able to warn inhabitants of at-risk homes to evacuate. The possibility exists using the NASA satellite system InSAR (Interferometric Synthetic Aperture Radar) that detects small movements on the ground. A study of an area in Louisiana, USA and noted that the ground shifted horizontally 10 in. (25 cm) in a section of the survey. A sinkhole opened up there a month later and the horizontal displacement observed was towards the center of the sinkhole [7]. Scientists believe that there is the potential to use this deformation as an early warning system in other sinkhole prone areas so that people can remediate the condition or evacuate buildings that might be at risk of a calamitous collapse. This is a step towards preparedness. However, they noted that not all sinkhole sites have surface shifting before a collapse, making necessary the previously cited methods to detect subsurface cavities that could become sinkholes.

Mitigation of an economic loss of home or business to a collapse into sinkhole event can be achieved by having collapse/sinkhole insurance as part of a home owners policy or an insurance policy that covers a business structure and inventory. The policies should cover collapse whether the event is from natural processes (e.g., subsurface dissolution of limestone) or from the failure of a water or sewer pipe and subsequent washout of supporting earth materials that leads to a collapse of an overlying home, business structure, road, highway, or bridge. Because many home buyers and home owners are unaware of their vulnerability (risk) from a collapse/sinkhole event, the state of Florida, USA mandates that home owners carry Catastrophic Ground Collapse Coverage. In the United States, Alabama, Kentucky, Missouri, Pennsylvania, Texas, and Tennessee are states with collapse into sinkhole problems and their own insurance requirements.

# 9.7   Health Hazards

## 9.7.1   Diseases

*Infectious/communicable diseases* are natural hazards abetted in some instances by human actions or inactions. Some of these diseases have been eradicated (smallpox) or nearly eliminated (polio, measles, guinea worm disease [dracunculiasis]). Others are treatable (malaria) or can be stabilized (HIV/AIDS), and still others are not preventable (dengue fever) but the exposure to which can be alleviated to some degree. Still others for which prevention and/or curing/stabilization medications are not available are being researched intensely such as an Ebola vaccine because of the 2014–2015 West African outbreak of the disease. In the most recent clinical trial of an Ebola vaccine, the success rate was 100 %, encouraging but requiring more testing before it is approved for universal application.

## 9.7.2   Mitigation

The threat or onset of infectious disease can be mitigated in several ways. First is prevention. There are vaccines available that can protect citizens against contracting specific infectious diseases. In the case of measles and polio, the actions of religious zealots (Taliban sect) who have burned down health clinics and injured or killed health workers have prevented vaccinations for all so as to achieve global eradication of poliomyelitis in two countries where it still occurs (Pakistan and Afghanistan) and measles (globally), especially in Asia and Africa. However, this not withstanding, vaccination programs for measles are making good progress. From 2000 to 2012 the number of deaths from measles has dropped 78 %, from 562,400 to 122,000. Mumps and measles (MMR vaccine) and whooping cough (DTaP vaccine) are two other infectious/communicable diseases that can be eliminated if vaccinations are given. In the case of children, a second application of the MMR vaccine is necessary to give them lasting immunity to mumps and measles.

It should be noted that vaccinations may not be close to 100 % effective. Scientists have modeled the probable effectiveness of vaccinations for endemic infections [8]. They conclude that vaccines denominated as "leaky" provide the same degree of resistance to a disease to all who have been vaccinated and gives a partial immunity after being vaccinated. A vaccine called "All-or-Nothing (AoN)" is best applied when the probability of re-infection is high, transmission is likely, or when a vaccine has low power to reduce the risk of infection. In contrast to a "leaky" vaccine, an AoN vaccine completely protects a major number of vaccinated persons but others in the population receive no direct benefit from it.

Some infectious/communicable diseases such as tuberculosis can be cured with medicines taken over a prescribed period of time. Chagas is a disease that is endemic in 21 Latin American nations. It can be cured if medication is taken early enough after the onset of the disease. If this does not happen the persons with Chagas disease can look forward to a middle age with cardiac and gastrointestinal problems with the healthcare burden this implies. Cholera is an infectious disease that can be controlled with access to adequate sanitation or cured by rehydrating victims with oral rehydration salts. If a cholera victim's dehydration is severe, IV plus antibiotics are used to cure the patient. A short term lasting cholera vaccine is available and used by health workers where there is a cholera epidemic.

AIDS/HIV is an infectious/communicable disease that is not curable and that has killed 40 million people. The disease can be stabilized if an infected individual has access to (or the funds to pay for) a cocktail of antiretroviral medications taken during a lifetime. This allows an HIV/AIDS carrier to live an otherwise healthy life and be productive in his/her community. About 15.8 million of the 36.9 million people with the disease are now taking the retroviral medications. Seventy percent of the global total of two million new cases are in Sub-Saharan Africa [9]. Unless the 21.1 million infected individuals not on antiretroviral medications get access to them and follow prescription protocols, the disease will continue spreading through populations.

Some infectious diseases can be controlled but not eliminated. Influenza is controlled by a seasonal vaccination that is prepared according to what scientists consider will be the dominant strains of the influenza virus during a coming influenza season. A strain was missed during the 2014–2015 season and more people suffered sickness as a result. Although some people do contract the disease, the vast majority of the vaccinated population is protected. If some do contract influenza from the unaccounted for strain, its effects are generally less than they would be without the vaccination.

The earlier cited Ebola epidemic that raged in Liberia, Sierra Leone, and Guinea, Western Africa controlled and the region declared free of the disease after more than 11,000 deaths of more than 27,000 infected. The Ebola epidemic shows how vulnerable many countries/regions are because of an inadequate health infrastructure that is not prepared to cope with a disease once identified, its spread, and the care and treatment of large numbers of infected people. Ebola is transmissible by body fluid contact and this had to be learned by family members and others caring for infected individuals as a first phase in controlling and ultimately leaving a country Ebola free. The spread of this disease to other countries, regions, or around the world was limited by the general immobility of people from afflicted countries and controls at transportation centers and by immigration controls at adjacent countries for those from the disease ridden countries who were mobile and presented a transmission risk. As the disease progressed in the afflicted nations, help arrived rapidly from developed nations that had experience in disease control that experts applied on site and taught to health givers in spite of the fact that WHO delayed before putting out a public health emergency alert. Laboratories that had been researching vaccines against Ebola increased their efforts and laboratory testing on animals. Test results of the promising vaccines on human subjects have been encouraging for two of the vaccines developed. Although the vaccines may prove to be effective against Ebola after future positive test results, they were not available at the time the epidemic was identified in West Africa.

### 9.7.3   Preparedness

Preparedness is the key to dealing with the outbreak of an infectious disease that could develop into an epidemic/pandemic. As noted above, initial preparedness such as protocols in the physical contact with afflicted persons that caused transmission of the disease, their treatment, and burial practices of their kin were lacking in the West African Ebola outbreak. A November, 2015 report by a panel from the Harvard Global Health Institute and the London School of Hygiene and Tropical Medicine critiqued the WHO for not declaring a public emergency until 5 months after being informed of the Ebola outbreak by Guinea and Sierra Leone [10]. The Panel recommended several ways in which the WHO could improve its role in preparing for and dealing with an infectious disease crisis. Among the recommendations were the following: (1) the need to invest in developing a nation's core capacity, that is, its ability to detect, report, and respond rapidly to outbreaks; (2) the strengthening of incentives for early reporting of outbreaks and science-based justifications for trade and travel restrictions to prevent transmission of a disease; (3) the creation of a WHO Center with adequate capacity to respond quickly to a disease outbreak; (4) an assurance of access to the benefits of research that yield improved diagnostics, and the most effective medicines and vaccines; (5) information on best protocols to follow in treating an infectious disease and on cultural awareness and traditions to account for when dealing with families of the afflicted or dead; and (6) availability of funding to put these and other recommendations into practice [10].

With respect to improved diagnostics (4 above), Nature published a December 2015 supplement in which contributors modeled the impact of new diagnostic and prognostic technologies for lessening the global burden of infectious diseases [11]. They are of the opinion that new diagnostics can more rapidly direct patient treatment and limit disease transmission to the general population thus effectively reducing the spread of epidemics. The effectiveness lies in the education of on-site health providers in the new diagnostics and protocols as they become operational. The contributors also believe that research can come up with new and rapid diagnostic protocols for multiple diseases that affect populations worldwide, that are effective and reduce costs that existing protocols incur, especially in less developed and developing countries.

The Ebola outbreak and spread by direct or indirect contact of an infectious disease raises the question of whether the world is prepared to combat an infectious disease, natural or from bioterrorism, that is transmitted through the air we breathe. Given the ease of transmission, the mobility of disease carriers within a country or internationally, the populated venues where transmission through respiration can take place (e.g., airplanes, cruise ships, theaters, arenas, subways, malls), the answer is no for many less developed and developing countries that do not have a well functioning health infrastructure... not enough doctors, nurses, well equipped clinics, hospitals, laboratories, medicines, or vaccines. Thus, a virulent disease transmitted through the air could infect and kill millions of people before medical care to treat it or vaccines and medicines to cure it are found and tested so as to control and/or eliminate a killer virus. It is possible to minimize the spread of an easily transmittable infectious disease but this requires a major investment. A proactive action would be for the health capable and economically advantaged nations to work with less prepared ones to create a good healthcare infrastructure, a mission of the WHO that has been neglected [10]. This would include training personnel as deemed necessary, building clinics where hospital care may be lacking, setting up laboratories that can identify diseases at their onset rather than waiting for results from a central laboratory, and by educational programs for citizenry. Response team training in all countries is essential to initially deal with a disease. Obviously, there has to be open and clear lines of instant communication between health organizations locally, regionally, nationally, and internationally when disease surveillance detects an outbreak node of a known or unidentified infectious disease. This allows a global response to begin on how to cope with the outbreak of a deadly viral disease in order to minimize its spread and develop a treatment protocol.

### 9.7.4  Diarrheal Diseases

Diarrheal disease that affect 100 s of millions of people each year can be controlled in two ways. First, they can be countered and their impact on society lessened by the use of antibiotics assuming medications are available and economically accessible. Second, is the identification and elimination of the sources of diarrheal diseases. Basically this means improving sanitation conditions and food storage, preparation, and handling operations. Investment by governments to provide access to advanced sanitation systems and for education with respect to food cleanliness can lessen the onset and spread of diarrheal infections and limit employee sick days, thus helping a country's productivity and economic development.

### 9.7.5  Vaccines in Research Phases

There are several infectious/communicable diseases that do not fall into the categories described in previous paragraphs. Vaccines and medications are being researched with some in trial stages. These diseases include malaria, dengue fever, yellow fever, West Nile virus, plague, and most recently Ebola because of the West African epidemic. As already noted, in the case of the Ebola epidemic, preliminary tests of new vaccines being researched have had good results on human volunteers. However, it is uncertain whether the positive results are because of the vaccines or the care factor given to those with the disease. A Zika vaccine is scheduled for Phase 1 clinical trials during the latter period of 2016.

Lastly it should be noted that Science Magazine commissioned a survey to prioritize ten potential vaccines that would merit increased government and industry funding for their research and development. These vaccines do not show clear scientific or safety obstacles and would benefit societal health conditions mainly in less developed and developing countries. The surveys were sent to 100 vaccine experts globally who were asked to prioritize the listed vaccines based on scientific feasibility, morbidity and mortality, and societal/economic impact. Fifty experts filled the survey. The overall priority ranking results in 2015 were (1) Ebola Sudan, (2) Chikungunya, (3) MERS, (4) Lassa fever, (5) Marburg, (6) Paratyphoid fever, (7) Schistosomiasis, (8) Rift Valley fever, (9) SARS, and (10) Hookworm [12]. Vaccines for some of these (1, 3, 4, 5, 8, and 9) could prevent or reduce the chances of serious outbreaks in the near future. Zika would be a 2016 addition.

## 9.8   Mental Stress

### 9.8.1   Mental Stress Evoked from Living Through Hazards

The psychological impact of living through a natural or anthropogenic hazard that injures or kills family and friends, and damages and destroys homes, business, and places of employment is a shock to citizens that puts them in a state of disbelief and distress. For that reason, the preparedness sections on hazards recommends that mental health professionals be available to the suffering populations. Mental health is considered here as a health hazard that in this text is triggered by a primary event but may result from a secondary event (e.g., earthquake aftershock) or a triggered hazard. The stresses caused can affect populations for reasons other than suffering physical, biological, and chemical dangers environments can present [13].

### 9.8.2   Principal Causes of Stress from Surviving a Killing, Destructive Hazard

Citizens living through a natural or anthropogenic hazard and secondary hazards that can occur suffer stress from several results of the event. The immediate stress factor is a concern for families in the degree of destruction citizens see wreaked on their environment. Stress levels heighten from the loss of or injury to family and friends and loss of home and possessions. The greater the loss, the higher the level of stress. Other factors that stress populations suffering through the immediate aftermath of a major hazard event include displacement, sense of vulnerability and insecurity, apparent slowness of assistance midst the chaos a hazard can cause, fear of what might come next, and physical exhaustion. There are ongoing discussions on coping with the stress of disaster [14, 15]. Post hazard stress arises from loss of employment and income, apparent slowness in recovery leading to societal normalcy, and reconstruction. In a latter section of this book, different classes of insurances are discussed that can, if purchased, mitigate economic stress from hazards losses. The number of people susceptible to hazard-caused stress and their resulting mental health problems will increase in the future because of population growth and increased population density in urban centers that have experienced recurring hazards. This is especially true for major cities (perhaps greater than one million inhabitants) and for mega-cities with more than ten million people. Easing or mitigation of stress levels involves having access to mental health professionals who can help relieve people's lasting anxieties, receiving social support from family and friends, focusing on the present and future rather than reliving the past, and taking care of oneself physically.

# References

1. Intergovernmental Panel on Climate Change (IPCC). (2014). *Climate change 2014. Synthesis report* (151 pp.). Geneva: IPCC.
2. Randerson, J. (2010). The shifting river that is making Uganda smaller. *The Guardian*, http://www.theguardian.com/environment/2010/dec07/climate-change-rerouting-semlike-river.
3. Zeni, L., Picarelli, L., Avolio, B., Coscetta, A., Papa, R., Zeni, G., et al. (2015). Brillouin optical time-domain analysis for geotechnical monitoring. *Journal of Rock Mechanics and Geotechnical Engineering, 7*, 458–462.
4. Allen, R. D. (1973). Subsidence, rebound and surface strain associated with oil producing operations, Long Beach, California. In D. E. Moran, J. E. Slosson, R. D. Stone, & C. A. Yelverton (Eds.), *Geology, seismicity, and environmental impact* (pp. 101–111). Los Angeles: Association of Engineering Geologists, Special Publication, University Publishing.
5. Farr, T. G., Jones, C., & Liu, Z. (2015). *Progress report: subsidence in the central valley, California* (34 pp.). Pasadena, CA: Jet Propulsion Laboratory, California Institute of Technology.
6. Bullock, P. J., & Dillman, A. (2003). *Sinkhole detection in Florida using GPR and CPT* (12 pp.). Tallahassee, FL: Florida Geological Survey.
7. Jones, C. E., & Blom, R. G. (2014). Bayou Corne, Louisiana sinkhole: precursory deformation measured by radar interferometry. *Geology, 42*, 111–114.
8. Ragonnet, R., Trauer, J. M., Denholm, J. T., Geard, M. H., & McBryde, E. S. (2015). Vaccination programs for endemic infections: modeling real versus apparent impacts of vaccines and infection characteristics. *Scientific Reports, 5*, 15468. doi:10.1038/srep15468.
9. World Health Organization. (2015). *HIV/AIDS fact sheet N360*. Geneva: World Health Organization.
10. Moon, S., Sridhar, D., Pate, M.A., Jha, A.K., Clinton, C., Delaunay, S., et al. (2015). Will Ebola change the game? Ten essential reforms before the next pandemic (18 pp.). *The Report*

*of the Harvard Global Health Institute and the London School of Hygiene and Tropical Medicine Panel on the Global Response to Ebola.*

11. Ghani, A. C., Burgess, D. H., Reynolds, A., & Rousseau, C. (2015). Expanding the role of diagnostic and prognostic tools for infectious diseases in resource-poor settings. *Nature, 528,* S50–S52. In Supplement. Infectious disease control and elimination: modeling the impact of improved diagnostics. doi:10.1038/nature16038.

12. Cohen, J. (2015). Vaccine priority survey. doi:10.1126/science.aae0168.

13. Murray, V., Aitsi-Selmi, A., & Blanchard, K. (2015). The role of public health within the United Nations Post-2015 framework for disaster reduction. *International Journal for Disaster Risk Science, 6,* 28–37.

14. Wiese, D. (2005). Impact of natural disasters on mental health. http://www.whitman.edu/live/katrina/wiese.pdf.

15. Mental Health America. (2015). Coping with the stress of natural hazards. http://www.mentalhealthamerica.net/conditions/coping-stress-natural-disaster. Accessed 2015.

# Chapter 10
# Magnified Impact of Natural Hazards from Global Warming/Climate Change and Inadequate Land-Use Planning

## 10.1 Global Warming/Climate Change

Global warming/climate change is a progressive happening that contributes to ongoing long-term disaster conditions and short-term hazards that put people and property at risk of injury, death, damage, and destruction [1]. The temperature increase correlates well with the ~40 % increase of $CO_2$ contents in the atmosphere from 280 ppm in latter part of the nineteenth century with the industrial revolution to the >400 ppm in 2015. The increase is attributed to industrial emissions of $CO_2$ especially from coal-burning electricity generating plants and cement manufacturing among others, especially from the 1950s and accelerated with increasing global industrial development into the twenty-first century. It is attributed as well to the loss of capacity of 'sinks' that absorb the $CO_2$ from the atmosphere (e.g., oceans and forests). The major sinks, oceans and seas with creatures that use $CO_2$ to precipitate their calcium carbonate ($CaCO_3$) shells, have become saturated to the point that their chemistries are changing with respect to acidity to the detriment of shelled forms in the marine ecosystems. In addition, another major sink, the areas of forests and other vegetation that use $CO_2$ for photosynthesis, are being reduced by forestry and human encroachment that has changed the natural land use patterns and helped lower the amount of $CO_2$ that could be absorbed by plants from the increasing amounts being emitted to the atmosphere.

Over the past two to three decades, the effects of global warming and the resulting changes in climate have manifested themselves worldwide in several ways that are sources of danger to life and property. In previous pages we discussed natural hazards and ways their impact on society may be mitigated. Climate change has abetted the risks some of these hazards and the causal effects they bring to populations worldwide. These include heat waves, drought, high energy storms (torrential, hurricanes, typhoons, monsoons; tornados, cyclones), and flooding. Meteorologists study historical records and modern happenings and report that the extreme weather driven events have increased in their frequency, intensity, and duration, likely added

© Springer International Publishing Switzerland 2016                                      53
F.R. Siegel, *Mitigation of Dangers from Natural and Anthropogenic Hazards*,
SpringerBriefs in Environmental Science, DOI 10.1007/978-3-319-38875-5_10

to by global warming and changes it is causing in world climate patterns. This increases the risk of injury, death, and property loss for populations plus infrastructure damage, whether in urban centers or in rural towns and villages. Climate change has expanded its warming reach into areas and altitudes from which it was previously absent. It is expanding the range for infectious disease spreading vectors (e.g., mosquitoes and malaria, dengue fever; deer and ticks) that threaten citizens, and for invasive life forms that attack food crops, thus threatening food security. As will be discussed in the following sections, only detailed planning can propel populations to actions that can mitigate the impacts of extreme weather conditions.

### 10.1.1    Sea Level Rise

Global warming has caused increased melting worldwide of mountain glaciers (e.g., in the Alps, Himalayas, Andes, Rocky Mountains), and ice sheets/shelves and continental glaciers (e.g., Arctic, Antarctica, Greenland). For example, Since 2012, the Zachariae isstrom glacier in Greenland is advancing at a rate of 2 km annually, with 2 km melting at its leading edge annually. It is expected to contribute to sea level rise for the next 20–30 years [2]. The contribution of melting from the glacial sources together with an increase in volume of seawater because of the warming and water expansion, has resulted in sea level rise that is ongoing. During the 100 years previous to 2015, sea level rose ~20 cm (8 in.) and by century's end sea level rise could increase by more than 50 cm (~20 in.) [1, 3]. The sea level rise threatens populations from encroachment that creeps over coastal areas and threatens to slowly submerge large areas of low topography islands and island states with elevations close to existing sea level (e.g., the Maldives). Rising sea level also abets erosion processes along coasts. Sea level rise intensifies the threats to life and property from global warming/climate change driven events cited previously such as recurring storm surges in coastal areas, flooding inland from torrential or long lasting tropical rains, and rarely from destructive tsunamis. In the future, global warming of sea water may affect ocean current flow that could result in changed weather patterns especially in the higher latitudes of the northern hemisphere. From 1971 to 2010, ocean waters warmed 0.44 °C in the upper 75 m. This is causing migration of temperature sensitive mobile fish and shellfish species from warming water to cooler waters (at higher latitudes) to the detriment of coastal villages that depend on the fish for food security. Increases in sea level cause an increase in water pressure on the continental slope. Where aquifers discharged at the slope whether as springs or where sea water already invades a fresh water aquifer, enhanced pressure can drive sea water landward possibly contaminating wells that provide citizens with fresh water for drinking, cooking, hygiene, and irrigation. As stated above, sea water rise is progressing and is projected to have a maximum increase of more than a half a meter (~20 in.) by the end of the century [1, 3]. This will increase the vulnerability from seasonal high energy tropical storms with their high velocity winds and storm surges, for low-lying coastal population centers, big and small, (e.g., Miami,

New York, Rangoon, London, Buenos Aires, Shanghai, Tokyo, and from North Sea storms that strike the Netherlands). Some researchers believe that even if the emission of $CO_2$ to the atmosphere ceased totally in 2015, glaciers will continue to melt and water volume continue to increase so that sea level rise will reach 3 ft (~1 m) higher at the end of the twenty-first century than it is today. They warn that planning and acting now for this future rise is essential.

Venice, Italy with its permanent population of more than 51,000 and a tourist population that can be 60,000 daily is sinking 1–2 mm annually, tilting slightly to the east towards the Adriatic Sea [4]. The subsidence is the result of compaction of underlying sediments and extraction of ground water and the slight tilt is likely the result of plate tectonics movement. The problem of flooding because of the sinking is exacerbated by sea level rise. One research group calculated that during the twentieth century the subsidence totaled 0.12 m and that sea level rose 0.11 m for a total displacement of 0.23 m [2]. Other scientists [5, 6] dispute some of the basis for the above conclusions [2]. They emphasize that their data show that the historic center of the city can be considered stable with a subsidence of <1 mm a year. It should be noted that the city is subject to having St. Mark's square under several centimeters of water (the author walked through on a visit) when there is a higher tide than usual. The situation can be exacerbated by sea level rise that increases the city's vulnerability to flooding but that may be mitigated by a barrier now being built to reduce the problem.

Mitigation can be achieved over many decades but only if there is a reduction in the emission of $CO_2$ and other greenhouse gases to the atmosphere, a buildup of physical defenses, vaccines or other health saving measures, and improved chemicals or genetic modifications that protect food crops. Given the politics of economic development, implementation of the former two methods seems unlikely unless 'realpolitik' takes hold and government decision makers realize that the increasing damage that sea level rise can do to many countries on the planet will directly affect many other nations as well.

### 10.1.2   Heat Waves

Heat waves cannot be prevented. Heat waves killed tens of thousands of European during extended periods of record-breaking temperatures, first in 2003 and then followed less than a decade later in Russia in 2010. The high heat over relatively short periods of time caused heat exhaustion and heat stroke among populations that were not prepared to deal with it.

Although heat waves and their temperatures and duration are not 100 % predictable, meteorological advisories can be announced so that people are aware of the probability of a heat wave and prepare to wait it out. Weather professionals follow high pressure systems at altitudes of 10,000–25,000 ft (~3050–7600 m). When the systems stall and strengthen over a region for a period of time, weather stations are alert to the possible onset of a heat wave. They surmise this because under high pressure, air moves towards the earth's surface and acts as a blanket that holds the atmosphere in and traps heat not allowing it to rise. Without the rise there is little or no convection of surface moisture upward and hence a lack of cloud formation and little chance of rain that can alleviate heat conditions. This leads to a buildup of heat at the surface that people sense as a heat wave. An IPCC report projects that as surface mean temperature increases over most land areas, it is likely that heat waves will occur with higher frequency and be longer lasting [1].

To lessen the chances of sickness or worse during heat waves, people should stay indoors out of the heat, keep hydrated, and keep cool either with electric fans or air-conditioning. This is easier said than done for much of the world population, especially for air conditioning, even in developed nations and for the more than two billion people without reliable or no electricity. Heat waves are especially dangerous for city populations because cities are heat islands. Roads, concrete, building rocks and bricks absorb the heat during the day and release it during the night lessening cooling that takes place. However, municipality preparedness can reduce the risk of sickness or death for its citizens. This includes checking on vulnerable citizens, especially the aged and families with small children, to make sure they are protected from the extreme heat. If this is not the case, police, firefighters, and civil defense groups should arrange transport to move them to cooling centers where they can find relief until the heat wave dissipates. Heat wave conditions can contribute to a toxic smog buildup, another threat to peoples' health.

### 10.1.3   Drought

Drought refers to times when there is less than the historically average rainfall (precipitation). They may be short term events or may last for several years. The shortfall may be minimal over a period of time or may be large and of great consequence to people especially if a severe drought lasts several years. This is because of its effects on a potable water supply and water for irrigation and livestock, hence food security. California, USA, was in the fourth year of a severe drought in 2015. The result was that the governor of the state ordered a 25 % reduction in use to conserve water resources in lakes, rivers, and aquifers. Only the economically important agricultural sector would initially be exempt from limits to water use but even here there was a conservation effort by many growers.

Many meteorologists have looked at historical and modern weather records and suggest that drought conditions in recent decades seem to occur more frequently and are longer lasting. Droughts can not be prevented but meteorologists can study weather patterns and give people alerts of the possibility of drought occurring or worsening and the possible duration of this natural global warming abetted disaster. As a result of global warming/climate change, the IPCC projects that there will be a reduction in renewable surface water and groundwater resources in most dry tropical regions, a condition that could lead to more severe droughts and increase the competition for water resources [1].

There are ways to prepare for and mitigate the effects of extended severe drought conditions. First would be to store water in surface and underground reservoirs when there is rain to be able to tap them as needed. Many communities that suffer recurring drought store a 3–6 months supply of water in such reservoirs. Second is to be able to bring in water from where it is plentiful either by tanker truck or by pipes pre-prepared to do so. However, this may be impractical depending on the infrastructure and the costs. Third is to encourage conservation or institute rationing of water resources. As noted above, this has been successful in cutting use by 25 % in California during the 2015 extended severe drought condition. Fourth, and costly to build, operate, and maintain, is to build desalination facilities if near an ocean or with access to a continuous supply of brine-laced groundwater. Finally, if clouds appear it is possible to seed them with silver iodide or calcium chloride crystals to bring moisture to a state of saturation that would result in rain.

### 10.1.4 Wildfires

Wildfires are both natural hazards started by lightening strikes, and hazard events started by human carelessness such as not being sure that campfires are completely out, by cigarettes tossed out of car/truck windows, by sparks ejected from wheels-track contact by speeding trains, and of course, arson. In some areas such as Southeast Australia and Western United States wildfires are recurring disasters. They are supported by high temperatures and long lasting droughts that dry out vegetation and thus facilitate combustion from a source cited above. There is no predictability as to where and when a wildfire will ignite, only that it can be a recurring event in some geographic regions. They don't often kill people but do burn out forests and brush vegetation causing natural disasters. In the 2015 summer, drought assisted wildfires in northwest United States (Washington State, Oregon and northern California) were disasters killing three firefighters. In northern California's Napa Valley, fires killed three people, caused the evacuation of 13,000 people, and burned down several hundred homes. However, there have been instances when many people were killed. Multiple wildfires in Victoria State, Australia during the austral summer, February 2009, with temperatures in Melbourne being over 109 °F (~43 °C) for 3 consecutive days, killed 173 people and destroyed thousands of structures.

Preparedness means having lookouts in forested/vegetated areas to spot the start of fires and to have firefighter crews and equipment ready to move quickly to contain them. Full containment of wildfires can take weeks or months. Wildfires are driven by winds and propagate by ejecting fiery embers significant distances as they spread. The spreading is also driven by updrafts in hilly regions. In theory, if homes are built in forested/vegetated regions, they should have vegetation cleared in a perimeter for at least 30 ft (~9 m) around them to improve their chances of surviving a wildfire. This is often impractical because of lot size or government dictum. A perimeter of 100 ft (~30 m) would be preferred. Other mitigation actions include having fire resistant roofing, removing lower branches close to the ground of trees taller than18 ft tall (~6 m), remove branches overhanging a roof or close to a chimney, remove dead leaves or needles from gutters, and have concrete driveways and gravel walkways around a home. Obviously, fast moving wildfires increase the need to evacuate citizens to schools and other public buildings in safe zones. Homeowners should have insurance policies with *full replacement liability* for fire losses.

### 10.1.5    High Energy Storms (Hurricanes, Typhoons, Monsoons; Tornados, Cyclones)

Meteorologists can forecast and monitor hurricanes (typhoons, monsoons). These hazards cannot be prevented. A review of storm activity over the past two to three decades, suggest that there has been an increased frequency, increased energy, and longer duration of these extreme events with time. This correlates with the increased rate of global warming in recent decades that has resulted in greater evaporation from a warmer and expanding reservoir of oceans as seawater volume has increased and sea water level has risen. Storms derive strength as they move across warmer ocean water and increase their moisture load. Meteorologists can track the storms, gauge their energy (wind speeds), model their probable paths and changes as they occur, and the probable rainfall they will deliver over an estimated period of time (torrential, sustained normal). They can also analyze the speed at which a storm is moving. For coastal areas, knowing wind direction, force, and tide conditions, meteorologists can estimate how far inland storm surges may reach. As such, citizens should receive timely advisories so that they can prepare for the extreme weather conditions to protect themselves and their property. In some cases, preparation is not possible in terms of evacuation to safe havens. For example, a storm with high energy winds (73 mph) is thought to be a main contributor to the demise of a cruise ship on the Yangtze river that cost 442 lives in June 2015. However, unsafe modifications to the ship, a lack of maintenance, and inadequate crew training and hence *lack of preparedness* for emergencies contributed to the sinking and loss of life according to a Chinese government official report. Historical records and modern data reveal that high energy storms driving extreme wave activity has sunk ships whether wooden under sail, or motor driven steel. The disaster events associated with sea level rise in addition to storm surges will be discussed in a following section.

Citizens can prepare for hurricanes (typhoons, monsoons) and lessen their impact at relatively little cost. For example, they can board up windows when they receive timely weather alerts. Windows could be blown out by high velocity winds. Certainly, having food, water, flashlights and portable radios and extra batteries, and a car charger for cellphones are essentials. A generator would be an asset if the storm damages infrastructure with a loss of electricity and people have to remain at home. Depending on their locations with respect to the path of a storm and its category (wind speed range), citizens may choose to evacuate to a safe location. Dangerous locations for people would include areas subject to flash flooding during torrential downpours or post-storm flooding, areas with hilly topography that could suffer landslides or mudflows during and after torrential rains, and coastal areas that could be reached by storm surges. After experiencing an extreme weather event and seeing what a storm has done to structures that were not constructed to withstand the wind velocities of the storm (e.g., tearing away roofing, siding), people can retrofit minimally scathed structures to withstand wind velocities that did the damage during a past event.

Meteorologists also monitor and forecast warnings on the approach of tornados (called cyclones in the southern hemisphere). These are land storms with violently rotating high winds that form a vortex that literally sucks up and can destroy an entire community or neighborhoods, tearing apart homes and infrastructure, and carrying off and dropping vehicles and livestock along their paths. Their swaths are generally limited but may rarely reach a mile or more across. As they move, tornados may literally skip along the earth's surface striking one population or building and missing the next. Tornados (cyclones) recur in the midwest and southeast of the United States and are common occurrences in countries in the western Pacific Ocean region (e.g., the Philippines, Australia). People escape injury or death by evacuating to storm cellars.

## 10.2   Hazards Triggered by Extreme Weather Events

### 10.2.1   Wind and Hail Damage

Sometimes overlooked in planning to deal with primary hazards from extreme weather events are the secondary or triggered hazards they cause. As noted earlier, these include heat waves and drought that can damage rain fed crops, threaten food security, and create conditions that support wildfires. High winds and large and small hail from major thunderstorm storms can damage or destroy vehicles, roofs, windows, landscaping and crops, and be an added threat to food security.

### 10.2.2   Flooding

Torrential storms and high energy storms can generate flooding, destabilize hillsides with water weight and lubricant waters that cause landslides and mudflows. As already discussed, they can generate storm surges together with high velocity winds that drive them and that can wreak damage and destruction in coastal zones and inland areas reached by sea water. As previously mentioned, where torrential rainfall causes sewer systems to overflow, there can be a release of pathogens into waters that are used for drinking and cooking, thereby causing sickness in the the consumer population. Most scientists worldwide are in agreement that global warming and the climate change it causes is a driving force for extreme weather events and their increased frequency, intensity, and duration.

# References

1. Intergovernmental Panel on Climate Change(IPCC). (2014). *Synthesis report*. Geneva, Switzerland: IPCC, 151 pp.
2. Graming, C. (2015). Rapid melting of Greenland glacier could rise sea level for decades. *Science*. doi:10.1126/science.aad7431.
3. Kopp, R. E., Kemp, A. C., Bittermann, K., Horton, B. P., Donnelly, J. P., Gehrels, R., et al. (2016). Temperature-driven global sea-level variability in the Common Era. *Proceedings of the National Academy of Sciences of the United States of America*. doi:10.1073/pnas.1517056113. 8 pp.
4. Bock, Y., Widowinsko, S., Ferretti, A., Novall, F., & Fumagalli, A. (2012). Recent subsidence of the Venice Lagoon from continuous GPS and interferometric synthetic radar. *Geochemistry, Geophysics, Geosystems*. doi:10.1029/2011GC003976.
5. Teatini, P., Strozzi, T., Tosi, L., Wegmuller, U., Werner, C., & Carbognin, L. (2007). Assessing short- and long-term displacements in the Venice coastland by synthetic aperture interferometric point target analysis. *Journal of Geophysical Research, 112*, F01012. doi:10.1029/200 6JF000656.
6. Teatini, P., Tosi, L., & Strozzi, T. (2012). Comment on Recent subsidence of the Venice Lagoon from continuous GPS and interferometric synthetic aperture radar by Bock, Y., Widowinski, S., Ferretti, A., Novall, F. and Fumagalli, A. *Geochemistry, Geophysics, Geosystems*. doi:10.1029/2012GC004191.

# Chapter 11
# Disease

## 11.1 Vectors Expanding Reach in Latitude and Altitude

Global warming and the resulting progressive change in climate zones has allowed the spread of vector-borne diseases into ecosystems that were cool and inhospitable to the vectors in the past but are now accepting of them as climates have become warmer and wetter. For example, mosquitoes have expanded their range to warmer higher latitudes and higher altitudes with the result that malaria and dengue fever have spread into regions that were previously free of them. The same is true of the tsetse fly and the encephalitis virus it carries, deer and the disease causing ticks they carry, and rodents bearing flea-borne disease viruses. If there is no mobility for vectors out of the regions where such diseases originate, the threats to other global locations are minimal. However, with mobility, vectors can carry diseases far distances. For example, in the middle ages rats bearing fleas that carried the plague arrived in Europe from the Orient on ships and originated a pandemic that caused the deaths of a third of the European population. In areas with increased rainfall and/or inadequate access to sanitation (e.g., regions in Africa and Asia), water-borne diseases will continue to subject hundreds of millions to diarrheal diseases and possibly epidemics from cholera and typhoid.

Disease to vegetation, including food crops, also spreads to areas that are warming and becoming more humid. This require the same control to safeguard vegetation as was used in areas where they are normally grown. This might mean the use of pesticides, herbicides, or plants that have been hybridized to resist disease by traditional hybridization or by genetic modification. Heat waves and/or extended drought that become more intense, more frequent, and longer lasting with global/warming/climate change are not properly diseases but can sicken and kill people, and wilt and kill crops and kill food animals by heat and dehydration, thus affecting public health and food security in many areas of the world.

© Springer International Publishing Switzerland 2016                                      61
F.R. Siegel, *Mitigation of Dangers from Natural and Anthropogenic Hazards*,
SpringerBriefs in Environmental Science, DOI 10.1007/978-3-319-38875-5_11

## 11.2   Humans as Vectors

People as disease vectors are a reality. If there is a human migration out of an area (e.g., in the search for water because of change in weather patterns, extended drought, or because of wars/conflicts) of one or more persons with a communicable disease, the disease can infect people in the receiving locale. This can possibly cause an epidemic, especially if a disease carrying person takes refuge in a densely populated urban center with limited access to medical facilities. Here we have as an example the recent spread of ebola in Western Africa that as of the end of 2015 was declared free of the disease. Ebola may no longer be a threat to humanity if the recent test of a vaccine that was reported to be 100 % effective against the disease withstands the rigors of additional clinical trials. There exists the possibility that MERS could become a threat from people who have travelled to or reside in Saudi Arabia and may unknowingly carry the MERS virus as they travel home or to other nations, but to the present, person to person transmission is difficult. From 2012 to August, 2015, in Saudi Arabia, MERS killed 483 of 1118 infected persons, a greater than 40 % mortality rate [1]. Initially, Saudi Arabian health officials did not report the disease to WHO as a potential epidemic. This was a mistake.

## Reference

1. World Health Organization (WHO). (2015). *Middle East respiratory syndrome coronavirus (MERS-CoV)*. Fact Sheet 401, June.

# Chapter 12
# Natural Processes Linked to Climate Changes that Threaten Food Security

## 12.1 Soil Salinization

Food security in many parts of the world is threatened when soil becomes non-productive or when arable land is lost. This can be considered a hazard that threatens the well being of citizens where salinization is an existing problem or could become a problem or intensify from global warming/climate change. In addition to a progressive loss of crop quality and yield from soil erosion and nutrient withdrawal without replenishment, the salinization process gradually coats the roots of growing produce with an accumulation crusts of salts that precipitate from rain or irrigation water until nutrients can not penetrate the crusts and plants wilt and die. Mitigation of this process is possible by regularly flushing out the salts before they reach concentrations that precipitate and encrust crop roots and moving the dissolved salt charged flush waters away from productive land [1].

## 12.2 Desertification

Desertification as defined by the United Nations Convention To Combat Desertification (UNCCD) is a class of land degradation that changes drylands (arid, semi-arid, and dry, sub-humid ecosystems) into deserts. Drylands comprise ~40 % of the Earth's land area (54 million km$^2$) Most of these drylands (58.5 %) are in Asia and Africa [2]. Desertification is a creeping hazard that is disastrous for vulnerable inhabitants. More than two billion people live in these regions and one billion are in areas at risk from the hazard. The United Nations reports that up to 50 million people could be displaced in a decade because of desertification [3]. Twelve (12) million hectares (>46,000 mi$^2$ or >121,000 km$^2$) of land are degraded by desertification, heat, and drought annually. Together with other land degradation processes (e.g., soil erosion, nutrient depletion), this threatened the livelihood of 20 % of the

© Springer International Publishing Switzerland 2016
F.R. Siegel, *Mitigation of Dangers from Natural and Anthropogenic Hazards*,
SpringerBriefs in Environmental Science, DOI 10.1007/978-3-319-38875-5_12

global population in 2015, food security for almost one billion people, and reduces a nation's productivity (GDP). Degradation globally is estimated to cost between U$S6.3 trillion and U$S10.6 trillion annually because of lost benefits such as from production of food, timber, pharmaceuticals, fresh water, recycling of nutrients, or absorption of greenhouse gases [4]. Desertification is a progressive process that is abetted by one or a combination of factors. These are the removal or loss of a vegetation cover by overgrazing, poor tilling of soil, and deforestation for fuel (e.g., wood to charcoal) and wood for construction (e.g., structure framing, flooring, furniture), and wildfires. The desertification process is likely supported by variations in climate that may be the result of global warming/climate change (e.g., drought, high temperatures).

The Food and Agricultural Organization projects that population growth alone (7.3 billion people in 2015) will increase the demand for food in 2050 (9.8 billion people) by 50 % [5]. Loss of agricultural land increases food insecurity and loss of forests and biodiversity is lost to expanding land suitable for agriculture. Climate change and erratic rainfall could cause a major decline of 25 % in the production of staples such as rice, maize (corn), and wheat that feed billions of people worldwide as arable lands contract and as populations grow in areas projected to be most affected by climate change (e.g., less developed and developing regions of Asia and Sub-Saharan Africa). Given resources and government will, soils degraded by desertification can be rehabilitated.

In theory and in practice, mitigation and reversal of desertification is possible if governments have the will to do so and if funding is available to support good farming practices.

Foremost might be to lessen erosion and evapotranspiration by winds by surrounding fields with vegetative windbreaks. This means investing in the restoration of forests, grasslands and other vegetation. Other practices include: (1) ban grazing or mandate controlled grazing; (2) adopt efficient irrigation systems for available water to crops (optimum amount of water delivered at times when irrigation is necessary = drip irrigation); (3) diversify crops to drought resistant ones; and (4) conserve soils and their productivity (limit erosion, replenish nutrients). In northwestern China, grazing and expansion of farmland are contributors to desertification. This is caused by socio-economic factors that are the dominant drivers that push economically disadvantaged farmers to expand their herds and farmland areas in order to survive [6]. This expansion has led to an increase in overgrazing and increases water use by crops that need more water than native vegetation that was stripped away for agriculture. To be successful, the mitigation practices cited above must be coupled with educational programs as deemed necessary and land tenure rights. This will give farmers the incentive to follow conservation plans (adjust to climate change) to preserve long-term crop productivity and hence their livelihoods.

# References

1. Siegel, F. R. (2008). *Demands of expanding populations and development planning*. Berlin/Heidelberg: Springer. 228 pp.
2. Sivikumar, M. V. K. (2007). Interactions between climate and desertification. *Agriculture and Forest Meteorology, 142*, 143–155.
3. United Nations. (2010). *Desertification*. Accessed at www.un.org/en/events/desertificationday/background.shtml
4. Doyle, A. (2015). Spread of deserts costs trillions, spurs migrants: study. Reuters, Sept 15.
5. Food and Agricultural Organization of the United Nations (FAO). (2012). *FAO statistical yearbook 2012*. Rome: World Food and Agriculture. 362 pp.
6. Feng, Q., Ma, H., Jiang, X., Wang, X., & Cao, S. (2015). What has caused desertification in China? *Science Report, 5*, 15998. doi:10.1038/srep15998.

# Chapter 13
# Land-Use Planning to Minimize Dangers to Citizens and Ecosystems

## 13.1 Planning to Protect People

There are hazard events that are caused by people and have an impact on them and their environment, impacts that can be intensified as direct results of human decisions or industrial/manufacturing activities. Many are related to shortcomings in land-use planning likely because the planners were not looking to future changes that could develop from the existing and progressing effects of global warming/climate change. In addition, there can be a lack of foresight on how global warming/climate change will affect growing populations and the need to sustain them at the least with food, water, and shelter. This unconscionable. Of great importance for growing populations is where to encourage or discourage human settlement. Such decisions should consider where to locate industrial facilities so that people and ecosystems would not be affected by their operations with respect to the wastes (emissions, effluents, solids) they would generate and how they would be dealt with as to capture and disposal. Such operations include electricity and heating generating facilities including nuclear facilities, chemical and pharmaceutical plants, mining and ore smelting projects, and agricultural projects including commercial food animal production sites and slaughtered food animal processing plants. The question of how to limit or stop the pollution these and other sources generate, either immediately or over time, was not foremost in the decisions made by early planners.

## 13.2 Avoiding Hidden Dangers

There are also hidden dangers from the past that can harm populations if they are not accounted for when expanding land use to accommodate population growth. There are unmarked buried disposal sites for containerized or loose toxic liquid and solid wastes as well as chemicals/raw materials in underground storage tanks from

© Springer International Publishing Switzerland 2016        67
F.R. Siegel, *Mitigation of Dangers from Natural and Anthropogenic Hazards*,
SpringerBriefs in Environmental Science, DOI 10.1007/978-3-319-38875-5_13

abandoned industrial and manufacturing operations. With time, metal containers can corrode and leak their contents that can contaminate groundwater or can rise to the surface and expose people to health-threatening toxins as was the case at Love Canal, New York, where, over time, a neighborhood was sited, unknowingly, over legally disposed of buried barrels of toxic chemical wastes. The wastes were slowly released when corrosion of the barrels allowed leakage and the escape and rise of the wastes to the surface contaminating homes, a school, a park, and other facilities as well as invading the groundwater system. This became the stimulus for the United States Superfund legislation that was the model for similar legislation in Western European nations.

Thorough land-use planning teams will use a community's master plan to show where new housing and infrastructure are to be located as population grows and where investment projects (potential sources of toxic wastes) that will offer employment will be sited. A search of government archives can also signal what ventures existed but were closed and their sites abandoned long ago. Records may show wastes they could have emitted, released as effluents, or buried without containment or as containerized liquid and solid wastes in areas now of interest for settlement. A search of newspaper archives can help reveal important information on past industries and the wastes they generated and disposed of. Former employees of the abandoned operations, or their families, may help reveal where toxic wastes were disposed of by burial. Once identified, the wastes must be removed before the terrain can be certified safe from contamination and be used for habitation.

Illegal dumping of urban and industrial toxic wastes is a major problem because without records or informants, the disposal sites are difficult to find. The wastes contaminate soils, surface and ground water and can affect air quality. A notable example of this exists in the area of Naples, Italy and provinces to the north and can be traced in many cases to illegal underworld (gangster) related dumping of toxic wastes ("black garbage") from northern and central Italy industries. This puts more than 500,000 people at risk. The toxic wastes that include arsenic, mercury, copper, cadmium, PCBs (polychlorinatedbiphenyls) and other hydrocarbons have likely caused increases in mortality of several diseases that have been reported on in scientific journals [1]. These include all cancers but especially on liver and lung cancer as long-term concerns and congenital anomalies as short-term concerns. Clearly, where illegal dumping exists or existed, land-use planning is problematic. Revelation of a possible problem may be indicated from health records that show nodes of toxic wastes related diseases in or near a location being assessed for development. Environmental scientists then seek to trace the source of a disease node that once found can be eliminated.

## 13.3   Response to Known Radioactive Waste Storage Threats

The Hanford Nuclear site covering 586 mi$^2$ in southeast Washington State, United States, is the most radioactive contaminated waste location in the western hemisphere. It is on the Columbia River. According to a U.S. Department of Energy May,

2015 report (www.hanford.gov), cleanup operations have been underway since 1989 with 82 mi² of active cleanup remaining. The site has been in operation since the 1940s producing plutonium for the atomic bombs that were dropped at Hiroshima and Nagasaki, Japan that effectively ended WWII. Subsequently, uranium fuel rods were produced for nine nuclear reactors along the Columbia River. Six of the reactors have been cocooned, two are in the process of being cocooned and one is being preserved. Associated facilities were demolished. The production processes produced solid, liquid, and gaseous radioactive wastes. Fifty-six (56) million gallons (204,400 m³) of high level radioactive wastes as liquid and sludge have been stored in 149 single shelled tanks and 28 double shelled tanks. Of the 177 storage tanks, 67 may have leaked and 6 giant underground tanks are definitely leaking toxic radioactive wastes. Thus far, pumpable liquids and 2 million gallons of solids (8810 m³) have been transferred from leaking single shelled tanks to double shelled tanks with 14 tanks retrieved. A waste treatment and isolation plant for the underground tank waste is 62 % complete as of May, 2015. Contractors project that the plant will be operational in 2022 with cleanup complete by 2040 but past projections have been missed by many years so that one can hope that the 2022 deadline will be met. The Hanford site encloses 1012 inactive dumps of which 939 have been remediated. The cleanup of buried or stored plutonium contaminated waste is more than 80 % complete. 15,000 m³ (almost 4 million gal) have been retrieved and shipped off site. There are 25 million ft³ (710,000 m³) of solid radioactive wastes in storage. Early in the project, contractors dumped or released radioactive wastes that directly contaminated air, soil, and water. Some radioactive wastes were dumped directly into the Columbia River. Of 200 mi² (518 km²) of groundwater contaminated by early contractors, 12.5 billion gallons (~48 million m³) were treated removing 157 tons of contaminants. As a result of the lack of control on the disposal of wastes by earlier contractors, 13,500 people living in towns downwind and down river of the Hanford Reservation may have been exposed to high levels of radiation that may have led to cancers and miscarriages and other health problems. In 1990, it was found that infants and children were subject to radiation poisoning from drinking milk…radioactive isotopes in the soil were taken up by forage being grown, eaten by dairy cows and transferred to milk. There was a report of a 400 % increase in rare birth defects (e.g., babies born without brains) in populations living near the leaking tanks.

## 13.4    Human/Ecosystem Protection in Planning for Massive Development Projects

### 13.4.1    International Road and Railway Projects in Africa

Land-use planning is of utmost importance on large scale development projects that cross a country or that are carried out in one country but affect the well being of the population and supporting ecosystems in a neighboring country. Similarly, joint regional projects that cross national borders can cause ecological disasters as they invade different ecosystems along their paths if thorough land-use planning among the countries participating is not done so as to meet project goals for all involved. Assessment of such massive development projects in Africa have been assessed [1]. An environmental team evaluated 33 development corridors to put in 53,000 km of roads and railroads ostensibly to increase food production to feed its growing populations. The roads and railroads will improve access to markets for farm products, allow farmers to efficiently bring in fertilizer and modern farm machinery, will make the transport of mineral resources and timber from source to delivery point easier, and give people easier access to healthcare and education. The corridors would likely include pipelines and power lines and attract human settlements. One such project would be a 4441 km (~2775 mi) for highway and railroad development from South Africa to the Congo. A West African corridor would run 4349 km (~2718 mi) from Dakar, Senegal to Port Harcourt, Nigeria. In both projects, the development corridors would traverse several countries where there are unique ecosystems that could be damaged such as equatorial savanna grasslands/woodlands, rain forests, and deserts. Analyses were made on a 50 km (~30 mi) wide band centered on a road or a railway for the 33 projects, 10 of which are now in progress with 23 in planning stages of which 9 are planned to upgrade infrastructure [2]. The analyses assessed the human populations in the study areas and estimated the agricultural production of each area's habitats that would be affected, and their environmental values. The latter were determined by the number of endangered species and native animals, vegetation diversity, critical wildlife habitats, and the capacity of the vegetation as a sink to absorb carbon dioxide (for photosynthesis) as a contribution to the efforts to slow global warming/ climate change. The results were reported "conservation value scores".

On the basis of the analyses of measured and observed data, environmental scientists reported that only 5 of the 33 projects met the goal of building in areas with high agricultural potential (lightly populated with suitable soils and climate) and less environmental value [2]. Six were classified as no good because they would invade areas with high environmental values without high agricultural pay back. Twenty-two were classified as marginal because they were good for either agricultural or the environmental, but not both. Of these 22, 10 projects have not yet started and the scientists think it important that they be governmentally and internationally reassessed. They emphasize how important it is for the future of our citizens and our planet to examine in detail, all facets of what massive development projects will mean to the populations and ecosystems in a country or countries involved as an essential phase of planning for the best use of land.

## 13.4.2   Proposed Large Scale Clean Renewable Electric Energy Projects for Africa

Land-use planning will have to be without error if African ambitions to increase the electricity generating capacity in the near future are achieved. At the 2015 climate summit in Paris, the African Union and the African Development Bank proposed a plan to add 300 GW (300 billion watts) with all clean and renewable electricity generating capacity to the continent's 2015 160 GW generating capacity. This is about half of Japan's electricity electricity production for a population that is a tenth of that of Africa. The added capacity would be with solar panels, wind farms, geothermal sources, and by taking advantage of the continent's massive water sources for generating hydropower. Of Africa's 2015 population of 1.17 billion people, 700 million Africans without electricity today would reap the benefits of electricity, mainly in Sub-Saharan Africa (~950 million people in 2015). The funding for this program would come from commitments from a US$80–90 billion annual fund economically advantaged nations expect to set up to help low income and lower middle income economies adapt to climate change. The World Bank will make its contribution via US$16 billion to pay for low carbon energy development for Africa. In addition, the European Union will contribute up to US$15 billion for clean energy projects. The installation of the additional 300 GW of electricity generating capacity is proposed to be done by 2030, that is, in 15 years. By that time, the African population is expected to grow to 1.64 billion people with 1.37 billion in Sub-Saharan Africa. Thus, if 700 million Africans are connected to electricity by that time, almost 500 million of the added population may lack this benefit. The picture becomes more complicated by 2050 when the African population is projected to be 2.47 billion of which 2.08 billion will live in Sub-Saharan Africa. Obviously, land-use planning to increase the continent's electricity generating capacity has to be done carefully to protect populations and ecosystems similar to what was done by in the study of large scale road and railway development projects in Africa discussed in previous paragraphs. This should be required by the funding groups. The land-use planning for these projects has to be thorough and complete especially with respect to the geology input on site selection for construction of hydroelectric dams and their effects on ecosystems and populations behind the dams and below the dams, and of the rocks comprising the valley wall as well as those underlying the dam site.(e.g., see examples of dam site problems in the following section). It is doubtful that the planning for, construction of, testing, and operation of hydroelectric projects and the infrastructure to support electricity transmission and distribution can come to fruition in 15 years. For such large scale clean and renewable energy projects, haste causes waste and cutting corners in any phase of a project to meet time and financial goals can mean disasters.

# References

1. Triassi, M., Alfano, R., Illario, M., Nardone, A., Caporale, O., & Montuori, P. (2015). Environmentalpollution from illegal waste disposal and health effects: A review on the "Triangle of Death". *International Journal of Environmental Research and Public Health, 12*, 1216–1236. doi:10.3390/ijerph120201216.
2. Laurance, W. F., Sloan, S., Weng, L., & Sayer, L. A. (2015). Estimating the environmental costs of Africa's massive "development corridors". *Current Biology, 25*(24):3202–3208. doi. 10.1016/j.cub.2015.10.046. Popular piece online at http://theconversation.com/massive-road-and-rail-projects-could-be-africa's/-greatest-environmental-challenge.51188

# Chapter 14
# Physical Siting of Locations for Housing, Commerce, Industry, and Parkland

## 14.1 Avoidance Siting

Where there is an existing population center in the a lower part of drainage basin, a consideration of new settlement upstream in the drainage basin has to evaluate its possible impact on the established population center downstream. Impacts could include a reduction of water supply or enhanced flood potential as vegetated/forested open areas in the upper reaches of the drainage basin are covered with housing and infrastructure so that water does not soak into the ground but rather runs off into streams/rivers that could cause flood problems in the down-flow areas.

Settlement in a floodplain is clearly not an allowable use of land. Similarly, settlement would be a bad decision in an area where topography, rock types (sedimentary rocks, especially shales/clay minerals and siltstones), geologic structure (beds dip towards the face of a slope rather than into a slope), and where there is a history of high volume, frequent and long duration precipitation that contributed to landslides and/or debris flows in the past. As stated earlier, these mass movements of earth materials can kill and injure people by impact and burial inside homes and destroy infrastructure that otherwise could help access in search, rescue, and recovery efforts. Where housing or businesses are proposed for an area with a sinkhole history, land use planners should require geological reports before construction permits are awarded to make sure that building plots are not underlain by underground voids that are subject to groundwater flow and ongoing rock (limestone) dissolution that can lead to collapse from weight of buildings or other stresses such as vibration and weight of vehicular traffic. It was previously noted that in western Pennsylvania, United States, there are abandoned coal mines that lack enough pillar support to prevent collapse. When the land was used for houses and infrastructure, there was stress on roof rocks from the weight of structures or roads built on them as well as from vibrations caused by nearby traffic. This resulted in structures and sections of roads subsiding or collapsing into the subsurface. In a 1996 publication, natural and anthropogenic hazards in planning development projects were discussed in some

© Springer International Publishing Switzerland 2016

F.R. Siegel, *Mitigation of Dangers from Natural and Anthropogenic Hazards*, SpringerBriefs in Environmental Science, DOI 10.1007/978-3-319-38875-5_14 detail so as to avoid events cited above and those that will be described in the following paragraphs [1].

In like manner, where there is active extraction of oil in an area or pumping of groundwater, there is a possibility of subsidence. Planning to put populations and infrastructure in such an area should include an assessment of the probability of subsidence that can cause property damage and disrupt infrastructure installations. This can be assessed by geological examination and study of well cuttings (core samples of underlying rocks) from which the fluids are being extracted. If subsidence is predicted to occur, either prevention methods have to be included in land use planning (e.g., recharge with fluids), or planners must revise plans for siting a population center away from the projected subsidence zone, or revise/disallow extraction projects. This latter option is unlikely to be selected because of the human need for water and the economic drive of nations for hydrocarbon sources of energy.

## 14.2   Dam Siting for Hydroelectric and Reservoirs Purposes: A Special Concern

Globally, there are 3700 large hydroelectric dams being built or with planning in progress for their construction. The majority of these are in developing nations with emerging economies in Asia, Africa, and South America, regions with large and growing populations [2]. The placement of dams for hydroelectric power, for flood control, to be reservoirs, and to provide for irrigation of proximate agricultural fields, has resulted in dam failures that caused injury, death, and property loss where planners have not done a detailed geologic study of a site selected for dam emplacement. A study of the geology as the first step in evaluating a potential dam site can avoid major disasters to people and ecosystems both downstream and upstream of their proposed locations and prevent losses for local and national economies. This means evaluating the geological history of the area for hazards that may have impacted a site in the past and could present problems in the future. It. means identifying the rocks in the valley and valley walls where a dam could be sited (igneous, sedimentary, metamorphic), and whether they might leak water into the subsurface through interconnected cracks and fissures in 'hard' rock or via porosity in 'soft' (sedimentary) rock and not hold water effectively. It means determining how the rocks in the valley walls will respond to water seeping into them as water rises behind the proposed dam to its highest level behind the dam. It means knowing the structural attributes of the valley's wall rocks as they may slope (dip) towards the valley or away from it. In addition, it means a knowledge of the type(s) of rock beneath the valley, proximity to faults in the subsurface, and how the subsurface rocks will respond to the stress put on them by weight of water held behind the dam. In a relatively recent disaster, water behind a dam seeped into sedimentary wall rocks that sloped towards a valley, adding weight, increasing pressure to push grains apart, reacting to cause sedimentary layers to swell and lubricating them leading to a massive landslide as described in the following paragraph. These are the same characteristics cited previously that cause landslides when rainwater or melting snow seeps into a hill.

### 14.2.1  Examples of Dam Failures/Disasters from Lack of Thorough Geological Analysis

In 1960, the Vaiont Dam in the Piave Valley, Italy was completed. It was 266 m (~870 ft) high. Geologically, the upper slopes in the valley are composed of limestones interbedded with montmorillonite shrink/swell clay rich shales sloping on one side towards the valley floor. During 1960, 1961, and 1962, as the reservoir filled, incipient land sliding was observed. This itself signaled that there was a serious problem that put populations at risk but additional incomplete research on slope stability indicated that the reservoir could be filled, a process that was completed in 1963. The water table behind the dam rose and infiltrated the slope adding weight, increasing hydrostatic pressure, and causing the clays to swell, all of which weakened the slope and this, together with in seeping from heavy rains, overcame the resistance to movement. The slope failed on October 9 at 10:39 pm when people were at home. The force of the landslide was such that it crossed the 99 m (~295 ft) wide valley and rode 135 m (~442 ft) up the opposite slope and displaced the water in the reservoir sending a more than 70 m (~229 ft) wall of water over the dam causing more than 2000 deaths downstream and extensively damaging property. Thorough examination of dam sites using geological observations and measurements and meteorological conditions became more meticulous worldwide after this signal event, but has not been perfect.

In another instance, at Koyna, India, a dam to serve as a reservoir was constructed in a seismically quiet area in 1963 with the lake behind it filled to capacity by 1965. There were known faults in the subsurface nearby. On December 11, 1967, a magnitude ~7.0 earthquake hit the region of the dam at 4:21 am killing at least 177 people, injuring 2300 more, and displacing 5000 people from 50 villages. The epicenter was near the vicinity of the dam. The earthquake was deemed by most scientists to have resulted from the weight of the water behind the dam where the reservoir was deep and extensive. This great stress on the underlying rocks, water seeping into the subsurface increasing water pressure from within possibly decreased frictional stress and hence shear stress of the rocks underlying the reservoir lake. The result was a reservoir induced earthquake [3, 4]. This is another signal event that should have alerted land-use planning teams worldwide of the potential of creating conditions that could cause damaging, sometimes killing, injuring, and destructive events where a dam is emplaced. However, as illustrated in the following paragraph, an alert may not be heeded with the depth of understanding that is necessary to make secure judgements about reservoir dam siting.

A site was selected for the construction of the Zipingpu Dam reservoir in Sichwan province China against the recommendation of scientists from the China Earthquake Bureau that, in 2000, warned the it would be built in the proximity of a major fault. The warning was not heeded because government inspections unspecified as to reliability, declared the area safe. In assessing the potential for reservoir induced seismicity, depth of the reservoir was most important with 100 m (~327 ft) being a safe depth, followed by the volume of water, and added that seismic responses can happen immediately after filling a reservoir or happen after a period of lag time [4]. The Zipingpu dam began filling during December, 2004 and the water rapidly reach a height of 120 m in 2 years. This was reported as time enough for water to penetrate deep into the earth's crust to decrease the friction between the fault walls. Water level in the reservoir declined from December, 2007 to May, 2008 indicating a developing problem. On May 12, 2008 there was the 7.9 Wenchuan earthquake that killed an estimated 80,000 citizens and damaged hundreds of dams as 300 km (~480 mi) of fault ruptured. The dam is 5.5 km from the epicenter. Scientists believe that the earthquake was the result of reservoir induced seismicity but are not making a strong pronouncement until they review data from Chinese scientists who had not yet made it available [5]. In a report on the social environmental effects of large dams, there is a good discussion of the relation between dams, failures, and earthquakes with a tabulation of reservoir induced changes in seismicity for dams worldwide from 1950 to 1979 including the Koyna, India event described above [6]. One would hope that reservoir induced seismicity as the cause of earthquakes is thoroughly evaluated by geologists and engineers before construction of high dams begins in seismically active sites such as in the Himalayas, Southwest China, Iran, Turkey, and Chile.

### 14.2.2   Some Human/Ecosystems Effects from Dam Emplacement

Land-use planners have to determine what effects dam placement will have in addition to those cited above. Large dams lead to an alteration of river flow that affects people and ecosystems upstream and downstream. Upstream, people are displaced from their homes and livelihoods and fertile lands are lost if the height of the dam backs up river levels so that centers of population, agricultural zones and sensitive, critical ecosystems would be inundated or starved of their normal water flow. To counter this effect, people can be resettled in homes equal to or better than homes that would be lost. The fertile lands that would be inundated would be lost and sensitive ecosystems could be disrupted, perhaps extinguished [2]. Millions of people were displaced and food productivity lost when China built the Three Gorges Dam, ostensively for flood control and hydroelectric power. As important was the purpose to bring river water to levels that would allow large ships to move upstream to more efficiently commercialize inland industrialized centers such as Chongqing. There is a question whether the resettlement policy China provided was fair, especially the land given to replace the land lost. The reduced flow caused by large dam emplacement can drain important wetland ecosystems downstream, change the sediment load delivered downstream that may destroy food fish spawning areas with the possible loss of species, disrupt natural flood cycles critical to downstream life forms, and depose a river from its natural flood plain. Decision makers have to determine if one or more of these conditions can be mitigated if changing the amount of water+sediment pass through from a dam can alleviate or eliminate the negative impacts on ecosystems. Thus, it is clear that value judgements have to be made to determine if a water project long-term benefits outweigh the costs to the environment. Obviously, where people are to be settled, experts have to evaluate any anthropogenic changes planned for the terrain that can affect citizens before action is taken to implement (construct) the changes.

# References

1. Siegel, F. R. (1996). *Natural and anthropogenic hazards in development planning.* Austin, TX: Academic and R.G. Landes Co.. 300 pp.
2. Zarfl, C., Lumsdon, A. E., Berkekamp, J., Tydecks, I., & Tochner, K. (2015). A global boom in hydropower dam construction. *Aquatic Sciences, 77,* 458–462.
3. Gupta, H. H. (1992). *Reservoir induced earthquakes.* New York: Elsevier. 364 pp.
4. Gupta, H. H. (2002). A review of recent studies of triggered earthquakes by artificial water reservoirs with special emphasis on earthquakes in Koyna, India. *Earth Science Reviews, 58,* 279–310.
5. Kerr, R. A., & Stone, R. (2009). A human trigger to the Great Quake at Sichwan. *Science, 323*(5912), 322.
6. Goldsmith, E., & Hildyard, N. (1984). Dam failures and earthquakes. In *The social environmental effects of large dams.* Cornwall, UK: Wadebridge Ecological Centre. Chapter 9. View at www.edwardgoldsmith.org/1020/dams-failures-and-earthquakes

# Chapter 15
# Pollution

## 15.1 Air

Anthropogenic pollution is a danger to otherwise healthy populations. Air may be polluted with potentially toxic heavy metals and fine-size particulates <2.5 μm, or gases/aerosols that when inhaled over time will cause sickness. Heavy metals (e.g., lead Pb, arsenic As, mercury Hg, cadmium Cd) or sulfuric acid ($H_2SO_4$) as aerosols ingested through respiration or deposited with rainfall can poison people and damage soils and thus harm food security. When food crops grown in polluted soils absorb one or more potentially toxic metals, they may not be edible because over-time, the toxin can bioaccumulate in consumers organs to health threatening concentrations. Mitigation can be achieved if sources of the dangerous emissions such as coal-fired power plants, smelters, battery factories and many other industries install and continually use and maintain emission control/capture equipment (chemical scrubbers and particulate precipitators). This can be mandatory and effective if government legislation demanding installation, use, and maintenance of such equipment is passed and the law enforced with unannounced visits by inspectors with the power to fine or close down a law-breaking operation. From a perspective of long-term benefits versus costs to reduce pollution, the health benefit to populations, societal stability, and uninterrupted work productivity will far outweigh the costs, use, and maintenance of the pollution control equipment.

### 15.1.1 Indoor Air Pollution

Long-term exposure to indoor air pollution in homes from cooking over open fires or using leaking stoves with wood, charcoal, dung crop wastes as the fuel for cooking and heating was linked the premature deaths of 3.3 millions people in 2012. This was mainly from the inhalation of <2.5 μm particulates, carbon monoxide, and

© Springer International Publishing Switzerland 2016
F.R. Siegel, *Mitigation of Dangers from Natural and Anthropogenic Hazards*,
SpringerBriefs in Environmental Science, DOI 10.1007/978-3-319-38875-5_15

other pollutants in gaseous phases. In decreasing order, the deaths were from stroke, ischaemic heart disease, chronic obstructing pulmonary disease (COPD), respiratory infection in children, and lung cancer. Much of the burden was in the Pacific Islands and Asia, with Southeast Asia, extending from east India to China, registering 1.7 million premature deaths attributed to indoor air pollution. Globally. WHO estimates that in 2012, there were 2.9 billion homes using wood, charcoal, dung and crop wastes as their main cooking fuel [1]. In theory, mitigation of this public health problem is simple: supply electricity or natural gas for household stoves. In practice, the funding to install the infrastructure to supply electricity or natural gas and the will of some governments to invest in the infrastructure are two limiting factors. In 2014, there were an estimated 2.5 billion people without access to a steady flow of electricity, more than a third of the global population, an estimated 1.2 billion people without access at all to electricity. Alternatively, improved stoves as part of clean cookstoves initiatives have been supplied to many households susceptible to premature deaths from indoor air pollution but these have often broken and are often used in conjunction with a home's inefficient air-polluting cookstove. Experts estimate that the total number of people using solid fuels in the growing populations in Asia and Africa will remain unchanged until 2030 and thus be a negative factor in achieving sustainable development [2].

### 15.1.2  Outdoor Air Pollution

During 2012, a estimated 3.7 million premature deaths were attributed to exposure to outdoor air pollution in cities and rural areas with 88 % in low and middle income counties mainly the Southeast Asia and Western Pacific regions [2]. The respired pollutants included <2.5 μm particles, gaseous and aerosol emissions downwind of power plants, smelters and other industries, burning of poor quality coal for heating in China, and near surface formation of smog. The <2.5 μm particles are especially dangerous because they are able to move deep into the lungs. The WHO set as a safe limit for these particles as a measure of air quality at 25 micrograms per cubic meter of air and this is greatly exceeded when a smog sets over an area. During the second week of November, 2015, with the cold weather and first snows in northeast China and the burning of coal for heating, the air pollution soared. This, together with input from industries caused the <2.5 μm particle concentration in 14 cities to rise above 300 micrograms per cubic meter and in three cities to be above the 500 level, extremely hazardous levels, 15 and 25 times the WHO safe concentration. At two monitoring stations the levels reached 1155 and 1400 micrograms per cubic meter, 46 and 56 times higher than the safe level. Health scientists report that 350,000–500,000 premature deaths annually in China are from outdoor air pollution and that if the Chinese <2.5 μm particle level safe standard were set at <40, this would mean 200,000 less premature deaths annually [3]. The deaths were linked to the same sources as for indoor pollution but the order changed so that stroke and ischaemic heart disease were followed in decreasing order by COPD, lung cancer, and respiratory infections in children. Particulate and gaseous/aerosol emissions can be mitigated by installation, use, and maintenance of different types of emission capture and control methods designed for specific emissions from different industrial sources.

Smog presents its own mitigation problems. Smog forms in topographic regions where a stagnant air condition prevents its dispersal by winds. This happens often in a valley or lowland surrounded by highlands. Smog occurs mainly where industrial and vehicular emissions react with ozone and sunlight as a photochemical catalyst to create a noxious gaseous mixture that can become health threatening to lethal for populations exposed to it for extended periods of time before weather conditions change from stagnant air to conditions that disperse the smog. The smog threat can be mitigated by preventing vehicular traffic from entering a population center under smog alert and by limiting or halting production at industries that emit the <2.5 μm particles and chemicals from which smog can develop until the smog disperses. This has been done effectively, for example, in the mega-cities Beijing and others in China, and in Mexico City. New Delhi, a mega-city with 16 million inhabitants and half as many vehicles, had the worst air pollution in the world in 2014. The New Delhi city government is trying to limit the problem starting in 2015 by allowing vehicle access on an every other day schedule. However, vehicles account for only 25 % of the <2.5 μm contaminants as do motor bikes and mopeds that also account for 25 % of the particulate pollutants but have no restrictions. Nonetheless, there must be industrial use of capture control technology and reduction in wood burning stoves in order to limit this toxin if some meaningful lessening of the New Delhi outdoor air pollution problem is to take hold. Limits on outdoor activity can reduce casualties from breathing air-borne toxins.

## 15.2 Waterways

Water may be polluted with potentially toxic heavy metals that may have come from atmospheric pollutants, from surface sources via industrial effluent runoff (e.g., arsenic As, cadmium Cd, mercury Hg, lead Pb), and from natural sources such as drainage over mineralized rock not rich enough in metals content to be mined. In addition, water bodies are polluted by acid rain, acid mine drainage, acid rock drainage, and nutrient-rich runoff from agricultural fields and animal husbandry. This can compromise a drinking water supply, irrigation waters (affecting food security), and be damaging to life forms in ecosystem waters, also affecting food security.

Arsenic (As) is a perfect example of a toxic metal foremost as a public health threat through drinking As contaminated water, cooking with it, and irrigating food crops with it. The threat is worldwide and has the special attention of health scientists in India and Bangladesh where millions of citizens are at risk of arsenic poisoning. Long-term exposure (ingestion), especially through groundwater, but also through food crops such as rice, leads to bioaccumulation and arsenic poisoning that can cause skin lesions and cancer of the bladder or lungs when it is present as the inorganic form arsenate ($As^{5+}$) at concentrations of 50–100 μg/l [4]. There is also an association with cardiovascular disease, neurotoxicity, and diabetes. Arsenic in contaminated waters may originate from multiple industrial sources and from the use of pesticides, animal feed additives, or pharmaceuticals. This health threat can be greatly mitigated by removing the arsenate at a groundwater well head or at a treatment facility before the water is used. One common method uses absorption by iron oxide and demands subsequent safe disposal of the sorbent matter. Iron oxyhydroxide impregnated in peat absorbed more than 90 % of arsenate in contaminated water tests in a 5 hours period [5]. Common anions in water such as sulfate, nitrate, and chloride had little to no influence on sorption onto the iron modified peat whereas phosphate and humid acid significantly lowered the sorption of arsenate. Other mitigation systems such as using phytoremediation by moving metal contaminated waters through a series of ponds where plants can absorb the pollutants, may be applied to metal contaminants in waters.

## 15.3  Soils

Soils suffer pollution by chemicals cited above and others from atmospheric and water sources that can render them unsuited for growing food crops and thus not productive. Soils can also suffer lack of productivity because of nutrient depletion. This latter threat to food security can be managed by the metered use of organic or chemical fertilizers. Soils may naturally contain pollution concentrations of potentially toxic metals. In this case, they are not suitable for agricultural plantings unless there is a crop that discriminates against toxic metals uptake whether naturally occurring or hybridized traditionally or by genetic manipulation to do so. There are technologies that can clean up contaminated soils so that they can be used for food crops but in most cases where the acreage (hectares) to clean up is large, thus is too costly to use. These technologies do not mitigate but rather eliminate the pollutant(s) problem if the pollutant source is basically eradicated. One such method is by excavation and is generally limited to <1 acre. The excavated soil can be treated at a facility to remove pollutants and then returned to fill the excavated site. Contaminated soil can be disposed of at a secure site. This and other physical, chemical and biological methods to remove or immobilize contaminants in soils, including phytoremediation (using plants to extract pollutants from soils) are used or being researched [6]. Again, to keep remediated soil safe for food crops, as forage for food animals, and for parks and playgrounds, the source(s) of soil pollution must be eliminated.

## 15.4  Oceans

### 15.4.1  Acidification

The oceans as a source of food are threatened by $CO_2$ pollution that is changing the water pH towards an acidic condition (acidification) but not yet close to being truly acidic. The oceans suffer dead zones from discharge of overloads of disposal of wastes to the seafloor that decompose by oxidation using all available oxygen (eutrophication) so that the oxygen depleted environment can not support marine life. Near shore ocean waters may receive nutrient-rich runoff from land that causes a rapid and far-reaching bloom of zooplankton. These have a short life cycle and can release neurotoxins into the marine waters when they die that leads to large scale fish kills, the so-called "red tide". Without reducing the $CO_2$ in the atmosphere, the drive towards ocean acidity will continue and ocean ecosystems will be disrupted and perhaps damaged beyond repair. Mitigation is possible if $CO_2$ emitted to the atmosphere is greatly reduced at industrial sources, and if vegetation sinks expand markedly. One can think outside-the-box and suggest outlandishly that billions/trillions of dollars be invested in setting up a long-term global network of billions of systems that pump the atmosphere into a potassium or calcium nitrate solution to precipitate $CO_2$ as potassium or calcium carbonate and return the $CO_2$ free air to the atmosphere. Given the small concentration of $CO_2$ in the atmosphere this will be a long-term continuous process that must be coupled with reduction of the emission of this gas at its multiple worldwide sources. Mitigation of nutrient runoff that feeds algae that can cause eutrophication and fish kills can be accomplished by enforced legislation that greatly reduces runoff from farm land and farm wastes into water bodies.

## 15.4.2   Overfishing

A human generated hazard threatens to disrupt ocean ecosystems and the oceans as a bastion of food security for tens of millions of residents of coastal villages and towns that depend on coastal fishing for their basic food and protein source. The Food and Agriculture Organization estimates that 2.9 billion people, mainly in Asia, get 20 % of their protein intake from food fish [7]. Billions more globally regularly consume food fish and shell fish from the open ocean. This hazard is overfishing to the point that the reproductive fish stocks can be reduced to near extinction levels. Overfishing is being countered with international and national laws designed to restore fish stocks in a reasonable time frame. The laws establish where fishing is permitted, the methods of capture used, and quota of tonnage of specific fish species that can be captured. Where laws are respected and enforced, this has allowed the fish stocks to recuperate in some fisheries [8, 9]. A serious obstacle to this is illegal, unregulated and unreported commercial fishing. This is being controlled to some degree by surveillance of national waters by national naval forces and by identifying and fining or closing the markets that have knowingly bought the illegal catches.

## 15.4.3   Oil Spills

Coastal populations and marine ecosystems suffer from oil spills that occur too often with the release of large and small volumes of oil into ocean waters from multiple sources. Some of these are discussed below in terms of mitigation possibilities and preparedness to limit the damage they can cause. Lessening the chance of marine oil spills from tankers that hit reefs (e.g., Exxon Valdez in Alaska) or run aground (e.g., Amoco Cadiz in France) as a result of human inattentiveness or electronic failures is possible by imposing and enforcing several requirements on owner companies. First is to assure that the shipmaster and crew are rested, clear headed, and alert. Second is to institute double hulling in newly constructed tankers with separate holding tanks for crude oil or refined products. If refitting existing tankers is feasible, it should be done. Third is to be certain that radar/sonar equipment is functioning as per specification and always attended. Lastly, it should be possible to install electronics that allow autonomous heading changes when radar/sonar feeds indicate an obstruction in a tanker's path if the obstruction is not detected by the bridge watch. Oil spills from tankers or from offshore drilling platforms that catch fire or suffer explosions represent less than 10 % of the contaminant discharged into ocean waters (>50 % from land via roadway runoff, leaks from industrial facilities including refineries and other sources). Oil spills receive most media attention when they are large and in coastal areas with sensitive ecosystems (e.g., habitats for marine life, birds, and furred mammals) and economic value (e.g., from tourism, food fish/shellfish habitats). Preparedness is possible to deal with the spills that can spread rapidly by having response teams that can deploy floating booms to contain a spill, use skimmers to capture oil, institute controlled burning, and introduce dispersants that can break the oil into small masses that are more easily degraded by weathering and the introduction of oil-feeding bacteria [10].

## 15.4.4  Radioactivity

Recently, with the 2011 destruction of the Fukushima nuclear power facility in northeast Japan, the escape of radioactive water into the Pacific Ocean is causing great concern as to how much radioactivity is bioaccumulated in fish stocks so as to make food fish inedible. This escape of radioactive water into the nearshore waters was ongoing in 2016 and is being closely monitored by the Japanese government and international agencies. The Tokyo Electric Power Company is installing barriers in the near shore to retain and capture the contaminated waters to prevent them from mingling with Pacific Ocean waters and is building a facility to decontaminate the radioactive water. In addition, a large area of residential living and farming is off limits because of radioactive fallout from the Fukushima event. Chernobyl in the former USSR is another prime example of an industrial accident that in 1986 released radioactivity into the environment that killed and poisoned citizens with radioactivity and also affected large swaths of nearby and far distant ecosystems via radioactive fallout.

# References

1. World Health Organization of the United Nations. (2014). *Household air pollution and health.* Fact Sheet 292, March.
2. World Bank. (2015). *Unlocking clean cooking and heating solutions key to reaching sustainable energy goals.* View at www.worldbank.org/en/news/feature/2015/05/19/
3. Chen, Z., Wang, J.-N., Ma, G.-X., & Zhang, S. (2013). China tackles the health effects of air pollution. *The Lancet, 382*(9909), 1959–1960.
4. World Health Organization of the United Nations. (2012). *Arsenic.* Fact Sheet 372, December
5. Ansone, L., Klavins, M., & Viksna, A. (2013). Arsenic removal using natural biomaterial-based sorbents. *Environmental Geochemistry and Health, 35*, 633–642.
6. Siegel, F. R. (2008). *Demands of expanding populations and development planning.* Berlin/Heidelberg: Springer. 228 pp.
7. Food and Agricultural Organization of the United Nations (FAO). (2012). *Statistical yearbook 2012.* Rome: World Food and Agriculture. 362 pp.
8. Worm, B., Hilborn, R., Baum, J. K., Branch, T. A., Collie, J. S., Costello, C., et al. (2009). Rebuilding global fisheries. *Science, 325*, 578–585.
9. Costello, C., Ovando, D., Hilborn, R., Gaines, S., Deschesnes, O., & Lester, S. (2012). Status and solutions for the world's unassessed fisheries. *Science, 338*, 517–520. doi:10.1120/science.1223389.
10. Water Encyclopedia. Oil spills: impact on the Ocean. Accessed 2015, from www.waterencyclopedia.com/Oc-Po/Oil-Spills-Impact-on-the-Ocean.html

# Chapter 16
# Mitigation of Hazard Impacts on Ecosystems and Their Inhabitants: Possible for Some, Not Possible for Others

## 16.1 Introduction

We have discussed the possibilities of mitigating the effects of impacts from natural and anthropogenic hazards on human populations and their physical environments. This is only part of the planet's environmental equation. We must also consider the impacts of hazards on ecosystems and life therein that is fundamental to human sustainability on earth. One can start with an assessment of the effects of global warming/climate change and the disruptions and disasters these effects can cause or support in ecosystems.

## 16.2 Migration of Life Forms

One result of global warming is that life forms that thrive in cooler terrestrial or marine waters migrate with the warming of their "natural" ecosystems to cooler ecosystems that they favor. In the oceans, for example, fish, some shellfish, and aquatic vegetation migrate to higher latitudes where the water is cooler or in some cases to deeper waters that have not warmed to the temperature that will spur their migration. Another result is that animals on land migrate from warming conditions in ecosystems to higher latitudes and high altitudes where cooler ecosystems are conducive to their needs. Where warming climates expand to higher latitudes and higher altitudes, insects (e.g., mosquitoes) and vegetation (e.g., invasive weeds) expand their ranges. The insect vectors can carry diseases, and invasive vegetation can have a deleterious effects on the habitats of life forms in the expanded ecosystem. Neither the migration of life forms from ecosystems that change from less hospitable to hospitable ones, nor the expansion of ranges of other life forms can be halted as global warming progresses. Neither is there expectation that the progressive warming can be reversed in the foreseeable future given the existing and likely

© Springer International Publishing Switzerland 2016
F.R. Siegel, *Mitigation of Dangers from Natural and Anthropogenic Hazards*,
SpringerBriefs in Environmental Science, DOI 10.1007/978-3-319-38875-5_16

future global political climate, not withstanding pledges made by many high emitting countries at several meetings over the past several years to 2015 to cut $CO_2$ emissions to 1990 levels in 10–20 years. India, the third greatest emitter of $CO_2$ is not likely to reduce its emissions because the government needs to support its industrialization and bring hundreds of millions of its citizens into employment and out of poverty.

## 16.3    Ecosystem Contraction

Another progressing problem caused by global warming is of ecosystem contraction as ice thins and melts in the Arctic, for example, reducing the "hunting ground" for polar bears and the Arctic fox that often feed on "leftovers" from polar bear kills. Already, polar bears have begun eating dolphin carcasses they find to replace of their traditional bearded seal prey as their snow/ice hunting areas are reduced. These arctic animals have no place to migrate to as their ecosystems contract to ever smaller areas. Time will tell if they can adapt to changing conditions in their habitat. Similarly, the encroachment into sensitive ecosystems by human endeavors reduces the areas of ecosystems for animals with the result that displaced animals with less area to serve their food needs invade the encroached zones now inhabited by people. These include deer, bear, wolves and other animals. Mitigation is possible to some degree for some of the impacts cited above.

## 16.4    Expansion of the Reach of Vector-Borne Diseases

The spread of vector-borne diseases such as malaria, dengue fever, and the Zika virus as mosquitoes expand to warming higher latitudes and altitudes can be mitigated by a population's use of insecticide, insecticide treated bed nets, and by insecticide spraying indoors and outdoors. As crop disease and invasive weeds spread to warmer climatic agricultural zones, herbicides or genetically modified crops can stem their negative effects on the food supply. Governments can preserve existing ecosystems and their life forms by prohibiting encroachment into forested terrain and other ecosystems by urban centers that need to accommodate growing populations, infrastructure, and industrial parks. However, many governments are loathe to do this because of their drive towards economic growth and its promise of combating poverty by industrializing, expanding employment opportunities, and increasing their tax base. Thus, given the important environmental impacts that cannot be arrested or otherwise mitigated, the human species has to adapt to conditions as they change and affect water and food security, safe sites for inhabitants, extreme weather conditions, and the spread of diseases, as they impact populations worldwide.

## 16.5   Global Warming Role in Coastal Hazards and in Ocean Acidification

### 16.5.1   Coastal Hazards Intensified by Global Warming

In addition to the increased volume of sea water from global warming discussed previously, warmer sea water has other effects that portent badly for the future. First, warmer sea water evaporates more moisture into the atmosphere. Clouds loaded with the moisture can carry it onto land where torrential rains can cause flooding. If flooding in an area is a recurring event from seasonal storms, the impacts can be meliorated as described in the natural hazards section if flood control is economically feasible for the at risk location(s). Also, when these torrential storms or high energy tropical storms (hurricanes, typhoons, monsoons) driven by high winds, and abetted by high tides assault coastal areas with storm surges, they can cause injury and death in a population and damage and destroy property and infrastructure farther inland with greater force. As with floods, economic investment in sea walls and bringing buildings to code can mute the effects of storm surges driven by high winds.

### 16.5.2   Effects on Marine Ecosystems in Addition to Fish Migration

For ecosystem populations, warm sea water can harm coral reefs, spawning sites for many important fish/shellfish species. If sea water temperatures rise 1 °C (1.8 °F) in a coral reef area, this kills off a symbiotic species, zooxanthellae, that provides food for the coral animal. The result is a bleaching of the coral that can lead to its death if the temperature rises to 2 °C (3.6 °F). As we learned earlier in the text, global warming is fueled in grand part by increasing contents of $CO_2$ in the atmosphere (baseline of 280 ppm in late nineteenth century to 400 ppm in 2015 with one or more ppm increase annually). As the partial pressure of the $CO_2$ in the atmosphere increases, more $CO_2$ dissolves in sea water. Sea water is not acidic but rather basic. Water that is neither acidic nor basic has a pH value of 7. A pH lower than 7 is acidic. A pH greater than 7 is basic. Seawater has a pH of about 8.1, a condition that allows animals to precipitate their shells efficiently from seawater. However, the continued addition of $CO_2$ in seawater as it builds up in the atmosphere has reduced the seawater pH at some locations to less than 8.1 but still greater than 8.0. At the lesser pH animals are not able to build their shells efficiently and have thinner shells. In some cases shells of existing animals can dissolve to some degree. Together, this can affect the oceanic spawning environments, the ocean food web, and can ultimately affect the global food fish supply. Without a dramatic change to a much lower $CO_2$ content in the atmosphere, an unlikely change for the foreseeable future, these damaging processes are increasing and for some, assuming a sense of is permanency . It is clear then that impacts of some hazards that global warming can bring to ecosystems can be mitigated to some degree whereas others cannot be lessened.

# Chapter 17
# Mitigation of Impacts on Ecosystems and Their Inhabitants Directly from Human Activities

## 17.1 Introduction

Human activities have polluted ecosystem atmospheres, waterways and soils. They have sickened and killed ecosystem life including human beings. This originates from coal burning power plants, smelters, and other industries that burn coal and use or manufacture chemicals, metal, and metal products, from agricultural chemicals runoff, and other sources. From the section on pollution we know that many of these sources generate emissions that include heavy metals, fine particulates ($<2.5$ μm), and gases (e.g., $SO_2$ that react with moisture catalyzed by the sun in the atmosphere to yield acid rain). Others generate polluted effluents that contaminate waterways and soils. In many countries, legislation passed to protect human health and the environment has required a great lessening of the emissions at their sources by the use of scrubbers that capture chemical emissions and precipitators that capture particles as they rise up chimneys. Laws also provide for treatment of effluents before discharge and/or a great reduction in their discharge. What has been lacking in some instances is enforcement of the legislation where plant managers limited use of available control and capture equipment or treatment protocols and/or did not maintain equipment to operate at maximum efficiency. This results in the use of less energy and reduced capital outlay to cut operational expenses and increase profits.

Countries with leaders that are educated to the negative benefits/losses relation that pollution brings to their lands and societal and economic well being, make and enforce laws to safeguard people and the ecosystems that help sustain them.

© Springer International Publishing Switzerland 2016
F.R. Siegel, *Mitigation of Dangers from Natural and Anthropogenic Hazards*,
SpringerBriefs in Environmental Science, DOI 10.1007/978-3-319-38875-5_17

## 17.2   Soil Damage from the Atmosphere: Mitigation and Reclamation

Soils repeatedly receiving acid rain become steadily less productive to non-productive because macro- and micro-nutrients are leached from them. This is in addition to the continued uptake of nutrients by growing crops where there is a lack of nutrient replenishment. The mitigation that can be effective in such cases is first by the capture of $SO_2$ emissions before they reach the atmosphere and produce acid rain that leaches soil nutrients, and second by the application of agricultural chemicals that replenish a soil's nutrient content. Both fixes have been adopted by most nations. Downwind of coal-burning power plants and smelters, heavy metals and other toxic chemicals comprising emissions that infiltrate soils when it rains can be taken up by crops to toxic concentrations so that crops are not suitable for consumption. Here, the mitigation is the same as has been previously described: use capture and control equipment (chemical scrubbers and particulate precipitators) where the toxins originate. The soils affected by heavy metals contents can be reclaimed by excavating them, subjecting them to high temperature combustion and putting the soil back in place but this is too costly for most agriculturalists. They can be reclaimed by using injected chemicals or added bacteria to mobilize the pollutants and move them out of the contaminated soils, or by growing plants that uptake large amounts of the metal pollutants either naturally or genetically engineered to do so but this is time consuming and expensive. Lastly, the polluted soils can be used to grow crops that discriminate against the uptake of the metal pollutants, again either by natural forms or forms genetically engineered to do so. This last method can be used where soils have natural high concentrations of potentially toxic metals if they are targeted to be used as agricultural fields.

## 17.3   Dangerous Disposal of Pollutant-Bearing Wastes

Solid pollutant bearing wastes disposed of at poorly designed surface disposal sites can react with infiltrated rain or surface runoff. A leaching reaction releases toxic components that can run off into proximate areas degrading soil or waterway eco-systems. The leachate may also seep into underlying aquifers and have a negative impact on groundwater ecosystems and on end users that pump groundwater for domestic and agricultural purposes. It may be possible but costly to rework old sites so that leachate can be captured and treated before interacting with ecosystems. This would involve capping them to prevent rainwater in seepage as one phase to prevent runoff that could disrupt proximate soil ecosystems and their potential productivity. A second phase to protect groundwater down flow from a disposal site is by putting in wells that can intercept the contaminated leachate and move it to treatment facilities before it intrudes the aquifer. This has been effective at the Love Canal site and other sites in the United States to prevent leachate from waste disposal sites from entering aquifer systems. The construction of new solid waste disposal sites where leachate is captured and moved to treatment facilities that detoxify them before their release to an environment is the ideal way to service urban centers with growing population. Certainly, the recycling of solid wastes such as paper/cardboard, plastics, glass, and metals is environmentally sound and if done right can be economically beneficial.

Ideally, polluted effluent from industrial operations (e.g., chemical plants), from agricultural runoff (e.g., from treated field crops and from animal husbandry), and other sources can be prevented on site. This is by capturing effluent bearing pollutants and treating them onsite or moving them to an off site treatment facility that removes pollutants before discharging the cleansed waters onto terrain or into waterways.

## 17.4 Special Sources of Fluid Wastes: Acid Mine Drainage and Acid Rock Drainage

Acid mine drainage (AMD) originates from water (rainfall, melting snow) moving through abandoned mines or through waste mine tailings. Acid rock drainage develops as water moves over or through rock that contains minor quantitates of ore minerals not worth processing. The AMD is physically difficult to stop and not an economically feasible undertaking. When the AMD enters ecosystems it is a continual process that kills most life it reaches. The process may last hundreds to thousands of years (e.g., new world mines or ancient Roman mines). Soils that absorb AMD lose their productivity and become, a type of brownfield. Waterways that receive the continual discharge of AMD can not sustain life until the acidity is reduced by dilution so that conditions are hospitable to life forms. Mitigation is possible if a costly investment is made to physically direct the AMD to holding tanks or ponds where they can be chemically neutralized before being released as safe, clean waters. However, this also is not often economically feasible. Where the AMD develops from waste rock tailings, the tailings can, in theory, be removed to areas where drainage through them can be captured and treated or they can be securely disposed of in deserts where groundwater is not threatened, populations are sparse, and lack of rainfall does not provide the medium for production of the AMD.

## 17.5 Tracking Down Sources of Contaminants That Invade Ecosystems

Environmental scientists can estimate where airborne and water borne pollutants originate or can trace them to their sources using principles of geochemical mineral exploration [1–3]. Dangerous impacts on people and ecosystems can develop from toxins that have been buried and thus hidden from view. They are traced, sometimes with difficulty, after health problems arise in a population, offending toxins identified, and their sources revealed after epidemiological studies and tracing to source by geologists/geochemists and support scientists and engineers. This includes toxic wastes disposed of in the oceans that damage marine ecosystems. Buried/hidden toxic waste disposal sites (e.g., with chemicals, radioactive materials) or leaking underground storage tanks (e.g., stored industrial chemicals, gasoline/diesel fuel) present hazards to ecosystems in soils, groundwater, and waterways where polluted water discharge. Containment and capture of the contaminated waters can mitigate the pollution problem but a complete solution is the removal of identified buried wastes, and treatment to non-toxic forms before release to ecosystems.

# References

1. Hawkes, H. E., & Webb, J. S. (1962). *Geochemistry in mineral exploration* (415 pp.). New York: Harper & Row.
2. Siegel, F. R. (1974). *Applied geochemistry* (353 pp.). New York: Wiley.
3. Rose, A. W., Hawkes, H. E., & Webb, J. S. (1979). *Geochemistry in mineral exploration* (2nd ed., 657 pp.). London/New York: Academic.

# Chapter 18
# Mitigation Economics

## 18.1 Funding Limitations/Sources for Mitigation Projects

In theory, mitigation is possible for many natural and anthropogenic hazards. In practice, however, the cost of mitigation to a good extent or to the maximum extent possible is great. It is beyond the means of many local, regional, and national governments in the low and lower middle income countries to support. Given adequate funding, the impacts of several natural and anthropogenic hazards and the events they may trigger can be mitigated to greater or lesser degrees, for example, by early warning systems (prediction), by barriers, by stringent enforced building codes (prevention). By bringing about a rapid as possible response to injury, sickness, and death (preparedness), and by the rapid repair of damage, reconstruction of destroyed facilities, and return to economic normality. However, the economic inequality among nations, and within national boundaries, plus a nation's priorities may prevent the adoption of very costly programs to minimize dangers to citizens and reduce property loss. In this case, there should be an efficient and prioritized use of resources that are available in order to minimize the dangers posed by a hazard. Indeed, the United Nations initially estimated that assisting lower income nations to mitigate the impacts for global warming alone would initially require US$100 billion with an additional US$400 billion necessary for full adaptation to global warming/climate change. The United Nations expects such funds to come from public and private sectors, bilateral and multilateral sources, and alternate sources of financing. The basis for the financing would be an international carbon tax (mainly from developed and selected developing nations), an international transportation and commerce tax, and a worldwide reduction in energy subsidies, a process currently being applied in many countries. For example, in Argentina, April 2014, there was a 20 % reduction in natural gas subsidies that saved the government US$1.6 billion. Similarly, in 2014/2015, Bangladesh followed an IMF mandate and slashed fuel subsidies that saved the country over US$600 million [1]. Overall, energy subsidies in 2013 totaled $548 billion with more than half of this sum to oil

© Springer International Publishing Switzerland 2016
F.R. Siegel, *Mitigation of Dangers from Natural and Anthropogenic Hazards*,
SpringerBriefs in Environmental Science, DOI 10.1007/978-3-319-38875-5_18

products. In lieu of parts of these sources or added to them, this writer believes that the United Nations should consider a global Mitigation/Adaptation Tax (MAT) to help low- and lower middle-income countries to adapt to global warming/climate change, similar to the VAT applied by many nations to generate funding for their programs. Those who spend more will pay a larger tax (mainly in developed or industrialized societies), and those who spend less will contribute to their own security by making a smaller but proportional contributions to help fund their own mitigation and adaption programs.

## 18.2  Benefits/Costs to Evaluate Targets for Investment in Mitigation Projects

### 18.2.1  Academic Exercise for Benefit/Cost Analysis

In an interesting academic evaluation of the best way to use funds to mitigate the impacts of natural disasters, evaluations were made on the expenses versus benefits of investing available funding to lessen an impact of earthquakes, of floods, and of high energy storms with data from 35 nations [2]. For earthquakes, researchers estimated that to retrofit all the schools in the 35 countries most prone to earthquakes would cost US$300 billion. The economic benefits would be far less than the costs. However, the investment would save more than 250,000 lives in the next 50 years. The cost would be highest in the most populated countries : China, more than US$100 billion; India, US$65 billion; and Mexico US$32 billion. For flooding, there were two options to mitigate the effects of flooding in 34 of the nations: (1) raise all buildings 1 m at a cost of US$5.2 trillion; or (2) build a community wall around affected communities as a structural risk reduction method that would save 61,000 lives over the next 50 years at a cost of US$940 billion. To cope with sea level rise of up to 1 m, these authors calculated that building a 1 m sea wall around vulnerable communities in the countries studied would cost US$75 billion but provide a long-term benefit of US$4.5 trillion that they would otherwise suffer from encroachment, erosion, and storm surges, a 60-fold long-term benefits to costs relation. In this scenario the long-term benefits would far exceed the costs. To reduce residential wind losses in 34 countries from high energy storms (hurricanes [typhoons, monsoons], and cyclones [tornados]), would cost US$951 billion for roof protection against wind related damage and would save 67,500 lives over the next 50 years and have a benefits to costs ratio greater than one as was not the case in the earthquake and floods examples. A valuation based on economic benefits to costs relation would opt for the investment that would reduce residential wind related losses if the building a 1 m wall to protect cities from sea level rise is not considered. But in terms of saving lives over the next 50 years, an investment to refit schools in earthquake prone regions would be the favored option. One factor that was not considered was how the estimated earnings over 50 years of the lives saved might affect the benefits to costs relations. This academic exercise is noteworthy but does not reflect the complexities of choosing or not choosing mitigation activities as does the publication cited in the following paragraph.


### 18.2.2  Real Life: One Dollar for Mitigation Yields Four Dollars Saved

One group of researchers determined that planners need objective evidence from several studies to show that under different environmental conditions such as geology, topography, and meteorology, a specific mitigation or more than one has been cost effective [3]. They cited three United States supported natural hazard mitigation grant programs that showed that for each one dollar spent for mitigation, society saved four dollars of future avoided expenses, a fine benefit/cost relation. Additionally, they emphasized that for reliable mitigation programs to reduce risk from a natural hazard and triggered events and by extension to anthropogenic hazards, decision makers and their consultant teams had to consider both the quantitative and qualitative losses and benefits to society and ecosystems locally, regionally, and internationally as required for specific projects .


## 18.3  Example of 'Not Costs Only' for Mitigation Project Decision Making

A rather thorough economic analysis of natural hazard mitigation projects that applies as well to anthropogenic hazards and hazards intensified by human activity was developed in 2011 by the Oregon Natural Hazards Workshop. The Workshop contributors followed 1998 FEMA guidelines with the purpose of reducing the impacts of hazard caused disasters in terms of lives saved, injuries minimized, and reduced or avoided property damage or destruction, and response, rehabilitation, and reconstruction costs [4, 5]. The Workshop Group emphasized the complexity of decision making on mitigation activities because they affect an entire community from individuals to public services and to businesses so that benefits and costs determinations are not always financially quantifiable. This is because of secondary happenings that affect social conditions and economic consequences such as job loss and reduction in tax revenues. For example, in a survey of businesses the Workshop Group found that 78 % of the respondents in Beaverton, Oregon listed the loss of electricity (and access to phone/internet) would be the most severe potential impact of a natural hazard disaster from an earthquake or a severe winter storm. In addition, 60 % of the businesses would suffer major financial losses after 8 days if pre-disaster conditions were not satisfactorily restored whereas 34 % would immediately suffer such losses. To cover themselves economically from hazard driven disasters, 54.3 %

of the businesses purchased insurance with 30.1 % purchasing business interruption insurance. Only 14.4 % had a business recovery plan. Lastly, less than 12.2 % of the businesses retrofitted structures to better withstand a disaster and less than 15.5 % made non-structural retrofits to save contents and inventories of their businesses. Such information was considered by the Workshop Group as they prepared an economic analysis of hazard mitigation projects.

The Workshop Group espouses three approaches for an economic analysis of potential mitigation projects. One is the benefits/costs relation that attempts to quantify whether the costs of mitigation before a hazard event brings financial savings benefits that lessen or avoid citizen deaths and injuries and property losses as well as response, rehabilitation, and reconstruction costs in the future if mitigation projects are not implemented. A second approach to decision making is a cost effectiveness analysis that prioritizes how best to use financial assets for mitigation projects. It assesses whether to put more funding into projects that protect citizens or property and businesses or try to establish multi objective mitigation activities. The third is the staples approach in which mitigation activity possibilities are assessed by constraints that exist that may be, for example, economic (e.g., funding available?), environmental (e.g., endangered species?), legal (e.g., within community master plan or not?), technical (limitations?), or social. The latter is with respect to acceptability and equity in the community population, and social disruption a selected mitigation project may cause. On this basis, potential mitigation activities may be ranked. If the perceived and measured future benefits of a dedicated or multi-objective mitigation project to a society are financially and socially greater than the "now" costs, operation, and maintenance, it (or they) may be adopted. To reiterate, this is with the full expectation that lives will be saved, injuries will be minimized, and property/business losses as well as response and recovery costs will be avoided or significantly reduced.

This underscores an important reality for decision makers. If a benefits/costs evaluation is made to determine if funds should be awarded for a mitigation project, it must be tempered by the socio-political benefits in addition to the economic benefits mitigation brings to a population. There is no question that disaster mitigation is cost effective [6]. It should be noted that there are less costly mitigation projects that are economically feasible for people to implement themselves if funds were available to purchase materials such as are used to reenforce roofs and boarding over windows for protection against high wind velocity storm. To be effective, such funding must be allotted wisely with reasonable fees for administration and without skimming by corrupt individuals entrusted to distribute funds as well as by those who receive funds.

# References

1. International Energy Agency. World Energy Organization Outlook 2014. www.worldenergy-outlook.org/resources/energysubsidies/. Accessed Nov 2015.

2. Kunreuther, H., & Michel-Kerjan, E. (2012). *Challenge paper: Natural disasters. Policy options for reducing losses from natural disasters: allocating $75 billion* (56 pp.). Philadelphia, PA: Center for Risk Management and Decision Processes, The Wharton School, University of Pennsylvania.
3. Godschalk, D. R., Rose, A., Mittler, E., Porter, K., & West, C. T. (2009). Estimating the value of foresight: Aggregate analysis of natural hazard mitigation benefits and costs. *Journal of Environmental Planning and Management, 52*(6), 739–756.
4. Oregon Natural Hazards Workgroup. (2011). Beaverton Natural Hazard Mitigation Plan. Appendix E., Economic Analysis of Natural Hazard Mitigation Projects, Beaverton, Oregon, 10 pp. Appendix C., Business Preparedness Plan, 16 pp. View at www.beavertonoregon.gov/index.aspx?NID=1166.
5. Federal Emergency Management Agency (FEMA). (1998). Report on costs and benefits of natural hazard mitigation. Publication 331, Washington, DC, 41 pp.
6. Kelman, I. (2014). Disaster mitigation is cost effective. Background note. World Development Report, CICERO, Norway, 4 pp.

# Chapter 19
# Insurance as a Mitigator of Post-Hazard Economic Stress: Coverage and Exclusions

Insurance against specific hazards is another way to mitigate their economic impacts and reduce mental stress for the insured. Swiss Reinsurance data in 2015 show that natural disasters globally caused an average of US$180 billion of economic damage annually in the last decade, 70 % of which was uninsured. This is because the costs for insurance, if it is available for a hazard zone, is beyond the economic reach of a majority of the citizens in many less developed and developing countries. Insurance companies are lobbying for the installation of defenses against natural disasters such as floods and adherence to up to date building codes in order to keep increases in premiums to a minimum and profits to a maximum...that is the way of business. Insurance premiums with respect to natural and anthropogenic hazards are based on risk assessment analyses discussed earlier in the text. The cost of insurance where it is available for a given hazard, with or without exclusions or additional endorsements, will vary greatly between locations. Whether people will be willing to insure against a hazard and pay for it depends on several factors. Principal among these is the experience a person has had with hazards, the person's property location with respect to a hazard zone and the reach of a hazard (e.g., of an earthquake, a flood, a wildfire). Additional among these factors are the codes according to which a structure was built, defense features, early warning systems (EWS), economic condition, and education (knowledge) about a hazard threat. An excellent reference to insurance for natural and anthropogenic disasters and secondary events is posted at http://www.insurance.ca.gov

© Springer International Publishing Switzerland 2016                                    101
F.R. Siegel, *Mitigation of Dangers from Natural and Anthropogenic Hazards*,
SpringerBriefs in Environmental Science, DOI 10.1007/978-3-319-38875-5_19

## 19.1   Homeowners and Automobile Policies in the United States

### 19.1.1   Homeowners Policies

Homeowners policies in the United States cover certain impacts from some disasters that damage or destroy properties but with specific caveats/exclusions. For example, most homeowners policies do not provide adequate coverage or exclude damage from hurricanes (typhoons, monsoons), high winds, and hail, and in high wind risk areas with high degree of wind damage. These policies exclude damage to property from earth movements including earthquakes, volcanic eruptions, tremors, landslides, mudslides, subsidence, or collapse. A State Farm Insurance Homeowners policy can cover property loss from the result of a volcanic blast, shockwaves, ash, dust (particulate matter), and lava flow. It also can cover damage from fire or an explosion that is caused by an eruption. In Hawaii, however, this insurance company will not write policies for property in high risk lava flow exclusion zones designated as 1 and 2 but will write policies in lesser risk zones 3, 4, and 5. *This emphasizes the importance for a property owner or prospective purchasers of other insurance policies to review each policy thoroughly before accepting it.*

Also, it is important to know that homeowners insurance (and automobile insurance) policies specifically do not cover losses when there is an act of war such as invasion, revolution, military coup, insurrection, strikes/riots or terrorism if terrorism is deemed to be state sponsored by the State Department. However, if a terrorist act is not shown to be state sponsored but rather carried out by a religious fanatic, fanatic group, or deranged individual, for example, homeowners policies cover damage caused by any resulting explosion or fire.

Insurance companies in countries other than the United States have their own rules and regulations, their own degree of coverage and exclusions, and their own fee structures. *Thus, it is absolutely essential that prospective purchasers of insurance policies read them carefully, perhaps with a lawyer to make sure that they are getting what they want in a policy.*

The sections that follow indicate what is covered or excluded from some policies that can be written to ease economic losses from natural and anthropogenic hazards. Where there are exclusions, citizens may opt to add such coverage as an endorsement to an existing policy at an increased premium or take out a separate policy to assure coverage.

#### 19.1.1.1   Wildfire Protection

Damage and destruction from wildfires is covered by homeowners insurance policies. This includes the building and personal possessions in the building. The policy also covers the cost of cleanup of dust and soot that accessed a building as the result of the wildfire and water/slurry damage caused by firefighting methods used to quell a blaze. To further protect a citizen, a policy should also include additional living expenses while a home is being repaired or rebuilt. Other structures attached to or on an insured property should have coverage of up to 10 % of a dwelling coverage. Values of buildings change yearly in many municipalities so that the insured value of a structure plus 25 % a homeowners policy adds may not cover all the rebuilding costs. *It is best to have a policy that pays the replacement cost of a burned down home.* Where one or more of these coverages is not included, it should be added as an endorsement to the homeowners policy. Policy premiums may be reduced if changes suggested by an insurance assessor are upgraded or added such as noted in the earlier discussion on wildfire mitigation methods or even the installation of a fire alarm or smoke detectors.

### 19.1.2  Automobile Coverage

Damage to vehicles from some natural and anthropogenic hazards (e.g., hail, fire) may be covered by automobile insurance when the policy includes comprehensive coverage. These are noted where applicable.

## 19.2  Tectonic Hazard Insurance

### 19.2.1  Earthquake Insurance

In the United States, earthquake insurance is most important in high risk areas such as California. Some citizens in other areas of the country that have suffered a major earthquake in the historical past or significant tremors more recently have opted to take out earthquake insurance. These areas include extensive parts of central United States, sparsely inhabited in the past but home to scores of millions now, areas rattled by the major La Madrid earthquakes in 1811 (magnitude 7.5) and 1812 (magnitude 7.7). A rare magnitude 5.9 earthquake at Mineral, Virginia in 2011, 83 miles southwest of Washington D.C., caused damage locally and in Washington D.C., notably at the National Cathedral and the Washington Monument. In addition, several residences suffered damage to chimneys and wall cracks. The earthquake was felt as far north as New York and as far south as the Carolinas, and as far west as Cleveland, Ohio. As a result earthquake insurance premiums for the Washington D.C. metropolitan area were greatly increased. Such rare earthquakes nationally prompt more citizens to take out earthquake insurance policies.

Coverage for earthquake insurance policies varies and carry a deductible that generally ranges from 2 to 20 %. Some polices do cover the repair of damaged buildings or replace destroyed properties, personal possessions lost inside a home when possible, and adjoining structures or sheds. An insurance premium varies with the provider. Locations that are close to an active fault and are historically especially prone to an earthquake have higher premiums as do older homes unless they have been retrofitted to higher code standards. Wooden homes withstand earthquake movements better than more rigid structures and have lower premiums. The type of soil or foundation a structure is built on also affects an insurance premium.

An earthquake insurance policy *should provide full replacement costs* for the home and possessions and living expenses for the insured if a home is destroyed and has to be replaced. This full replacement feature can be purchased either as an endorsement on a homeowners policy or as a separate policy. The policy should be carefully reviewed for the deductible required and any exclusions or limitations that might cause problems for the insured. For example, the California Earthquake Authority allows a 10–15 % deductible for participants, insures structures with up to four units, mobile homes, condominiums, and townhouses, but does not insure commercial, industrial, or business properties.

There is a suggestion that for low-income property owners in a developing country, there can be a voluntary collective earthquake insurance policy financed by a property tax assessment [1]. This is a cross subsidy from economically advantaged home owners and low-income home owners. It promotes an insurance culture while strengthening community relations. Both pay and both benefit in the event of a damaging/destructive earthquake.

A pilot program for earthquake insurance in rural China found that 88 % of the 681 people polled were willing to buy the insurance and pay a 160 Yuan annual premium (~US$25) based on risk perception (experience) and risk exposure (location, strength of structure) [2]. Those people living in reinforced houses were less willing to pay but economically advantaged people were more willing to pay partly because of government propaganda for earthquake mitigation.

### 19.2.2    Volcano Insurance

In the United States there are six states with active volcanoes: Alaska, California, Hawaii, Oregon, Washington, and Wyoming. Volcano insurance may or may not be available depending on the risk zones determined from past eruptions and from the characteristics of past eruptions (e.g., ash/tephra emission and fall, lava flows, volcanic mudflows/lahars). Coverages for a specific volcanic hazard may not be covered. As cited in the homeowner's policy coverage, most State Farm Insurance Homeowners policies cover property loss from the result of a volcanic blast, shockwaves, ash dust (particle matter), and lava flow. They also cover damage from fire or an explosion that is caused by an eruption. In Hawaii, State Farm will not write policies for property in mapped high risk lava flow exclusion zones but will write policies in lesser risk zones. Damage to vehicles from volcanic activity may be covered by automobile insurance when the policy includes comprehensive coverage.

Homeowners policies exclude damage to property from earth movements including earthquakes, tremors, landslides, mudslides, subsidence, or collapse whether caused by a volcano erupting or not. The cost to remove ash that does not cause a direct physical loss to a property is likewise not covered by a State Farm Homeowners policy. To cover such a happening one needs to pay a premium for an endorsement or purchase a separate policy. Other exclusions exist so that *each policy has to be reviewed thoroughly by a property owner*. For example, the Washington State Office of the Insurance Commissioner advises Washington State Homeowner policy holders that in addition to covering property damage from ash, dust, particle matter, and lava flow, the policy covers the cost of removal of ash, dust, and particle matter from the interior and exterior of a home. What is excluded from Washington State Homeowners policies are later ash falls caused by winds following an initial eruption, earth movements, tremors caused by an eruption, landslides, and shockwaves. This exclusion originated from the Mount St. Helens 1980 eruption where shockwaves from the lateral blast resulted in extensive ecosystem damage. State Farm policies do include damage to property from shockwaves.

## 19.3   Flood Insurance

Few inhabited locations on earth are free of the threat of flooding. This hazard originates as the result of one or a combination of events, natural and anthropogenic. As noted previously in the text, the natural events include hurricanes, other severe torrential rainstorms, rising water levels near settlements, rapidly melting snow packs, overtopped or breached levees, and tidal surges. The human factors that abet flooding in some locations include the installation of flood control measures in other areas (e.g., flooding upriver areas from dam building downriver), and clogged drainage systems or old systems that cannot carry away an increased water flow (from imperfectly planned land use changes). The land use changes come with development such as cutting trees [deforestation], building subdivisions with attendant infrastructure such as paving roadways and sidewalks with impermeable surfaces that replace the soil cover and natural systems that can soak up heavy rainfall (or snow melt) and reduce runoff into stream or river channels.

The United States has a National Flood Insurance Program (NFIP) that was created to protect families from financial distress because of damage from floods. For this program, an area is considered flooded when at least two acres of otherwise dry land is submerged or if two or more properties are inundated. Mudflows triggered by excessive rainfall or snowmelt fall under NFIP policies. Not all sites are eligible for coverage. The Program covers only those locations where flood control measures exist. The Program insurance is available through licensed companies. Coverage for a home can be up to US$250,000 and for personal belongings up to US$100,000. Premiums for the same property vary among companies so it is prudent to get multiple quotes. Insurance is available for dwellings in the low and moderate risk flood zones. Premiums will vary with where a property is located physically with respect to flood risk zones mapped by geologists and hydrologic/hydraulic engineers specialized in flood hazard mapping. The zones include the 100 year flood plain (high risk zone with one chance in a 100 for flooding, an estimate that is often exceeded), on up to the 500 year flood plain beyond which is a low, preferred risk zone. To determine the zones, scientists use topography and elevation data, watershed (drainage basin) flow history, coastal areas evaluation, rainfall and storm patterns, and storm surge data. Depending on the zone a property is in, premiums will differ. For example, the cost for insurance in a preferred risk zone might be US$317 annually for US$250,000 for a structure and US$196 for US$100,000 for personal belongings if there has been no previous flood insurance claim. If there was a previous claim, the premiums would rise significantly. In a higher risk location the premiums may be US$2400 annually and US$1000, for the structure and personal belongings coverage, respectively. As with other types of insurance, selection of a high deductible can lower premiums. Properties located in coastal zones inland of a given elevation above the high water elevation and where there is a history of storm surges may have high premiums such as US$5000 annually to buy US$250,000 for structures and US$2200 to buy US$100,000 for personal belongings. In the United States the average flood insurance claim in 2012 was US$38,000. When purchasing a home or other property (e.g., farmland), it is important to determine if the home or land has a history of or an insurance claim for flooding.

## 19.4   Insurance Against Mass Movements

### 19.4.1   Against Landslide Damage/Destruction

Generally speaking, there are few insurance companies in a few states in the U.S. that offers policies that cover landslides or the other damaging earth movement hazards collapse and subsidence. One exception is the possibility of getting a landslide policy issued by Lloyds of London is for California homeowners. However the Lloyds policy is available only in neighborhoods where there has been no previous landslide damage and then a home must not be on a hill (slope) or beneath it, must be on firm soil, and lastly must have a good drainage system. The cost is US$0.40 or US$400 for a $100,000 policy with a deductible of 2.5%. One insurer wrote that under these conditions landslide insurance would not seem to be necessary. Some insurers will issue a "change in condition" policy to cover landslides. Damage from rockfalls is generally covered by a homeowners insurance policy under a falling object clause.

### 19.4.2   Against Collapse/Sinkhole Loss

As discussed in the earlier section collapse/sinkhole, damage occurs when there is a ground cover collapse. This most often occurs in areas underlain by limestone in the subsurface where limestone has been dissolved over millions of years by groundwater leaving large voids beneath the earth's surface or in areas where there are abandoned coal mines. Mortgage companies usually require sinkhole insurance as part of a homeowners policy, especially for homes purchased in areas where known sinkhole damage has occurred or is likely to occur. In the United States. This would be, for example, in Florida (state with the biggest sinkhole problem), Kentucky, and Tennessee where sinkholes are in limestone areas, and in Pennsylvania where sinkholes are in areas underlain by abandoned coal mines. To collect on a policy, a property must be damaged or destroyed by the actual sinkhole activity. As with all sinkhole insurance policies as part of a homeowners policy or a separate one, *it is necessary to carefully read the policy* to determine what is covered, the deductible responsibility, the premium, and what is excluded.

### 19.4.3 Against Subsidence Damage

Subsidence is the downward movement of a site on which a building stands where the movement is not connected to the weight of the building. The site would subside even if there was no building on it. The subsurface is subject to compaction and thus unstable. As noted earlier in the text, subsidence is most often the result of the withdrawal of fluid (water, oil) from the subsurface. If subsurface rocks do compact, the rocks and soil above it and the surface may subside. Buildings or infrastructure in an area suffering subsidence will be damaged (e.g., water and natural gas lines, roads). Subsidence insurance may be purchased as an endorsement to a homeowners policy. Each municipality has its own norms so that *a policy must be reviewed carefully* before purchase. Since 1999, subsidence of houses sited over worked out coal mines in Pennsylvania, U.S.A. has been covered by the state with the Pennsylvania Mine Subsidence Insurance Fund at a cost to the home owner of US$150 for US$250,000 of coverage.

## 19.5 Agricultural (Crop/Livestock) Insurance

Agricultural insurance can cover crop and/or livestock loss mainly from floods, hail, windstorms (including tornados/cyclones), drought, and pestilence. Individuals select the crop or crops and/or livestock they want to protect from a specific weather hazard or a combination of weather hazards. This insurance is important to nations globally because its helps to maintain food security for the world's growing populations by keeping farms operational so that farmers can replant crops and/or restock livestock that was lost. Agricultural insurance schemes were reviewed for the European Commission [3]. One problem is that penetration of the market has been difficult, especially in developing nations but is increasing because of government subsidies to help cover the cost of premiums [4, 5]. The percentage of the market premiums written range from 62 % in the United States and Canada to 18 % in Asia, 17 % in Europe, 2 % in Latin America and 1 % in Africa. The status of agricultural insurance in Latin America and that in Asia and the Pacific region has been reported on [6, 7]. The publications cited above are recent and attest to the importance the Food and Agriculture Organization of the United Nations and the World Bank give to the need for agricultural insurance especially in less developed and developing nations where population growth is the highest.

It is interesting to note that China has been testing crop and livestock (breeding-pigs) insurance against natural disasters since 1949. Different models were tried but were not successful until 2007. In 2007 a model was adopted with insurers being commercial insurance companies competing in a market-driven economy and with subsidies from the central and provincial governments that generally were greater than 50 % of the premiums. The insurance covers the basic costs for seeds, fertilizers, pesticides, irrigation, machinery, and mulching film, against loss from rainstorms, floods, water logging, windstorms, hail, ice storms, and drought. It also covers certain diseases in crops and livestock. Premiums are at 3–10 % of the insured amount with this amount set by the companies. Because most losses experienced came from flooding and water logging many farmers believed that funds used for subsidies should be put into disaster defenses and mitigation such as flood control methods. In Hunan Province, for examples 58 % of the farmers supported this idea while 24 % supported the subsidies [8]. An interesting paper reports on the state of government subsidies for crop and livestock insurance [5]. It also shows government subsidies as a percentage of premiums paid by producers in selected countries in 2007, and gives options for developing countries.

## 19.6    Drought Insurance

### 19.6.1    For Other than Crop/Livestock Loss Protection

Drought insurance coverage for other than agricultural losses can be purchased to protect homeowners against the cost of foundation repair. In this case, under severe and prolonged drought, soil dries out, shrinks, contracts and pulls away from a house and weakens the foundation causing cracks and also can cause pipes to crack, leak or even burst causing extensive damage and put the burden of repair on the homeowner. This is not covered by homeowners insurance. If a homeowner has an *all perils* rider on the homeowners policy the damage from a prolonged drought is covered. This potential foundation problem may be mitigated by watering the soil 1–2 ft away from the hours every day but during a severe drought it would be unwise or unlawful to use water for this purpose.

## 19.6.2   Associated Heat Waves

Heat waves may be associated with drought but there is no heat wave insurance. People can protect themselves with air conditioning, ceiling or floor/table fans, water compresses on back of neck and wrists, showers, blasts of cold water from spray bottle, and by shutting off sources of heat. People should recognize when heat exposure is of concern if they develop a heat rash, have cramps, are becoming exhausted, and may be at the start of heat stroke. Treatment for these conditions is covered by medical insurance. If there are wildfires caused by or abetted by heat waves, the damage or destruction suffered should be covered in the homeowners insurance policy.

## 19.7   Extreme Weather Event Insurance

### 19.7.1   Against Windstorm Damage

Most homeowners policies do not provide adequate coverage or exclude damage from hurricanes (typhoons, monsoons), high winds, and hail, and in high wind risk areas with high degree of wind damage. One needs a separate policy with a separate premium. This policy may require notification of a hurricane watch by weather bureaus that winds be greater than 74 mph (119 km/h) at the specific location. A premium may be reduced (up to 30 %) by a positive report after a Wind Mitigation Inspection. This includes a verification of construction methods such as roof shape, reinforced roof decking with longer nails or screws, the roof shingle attachment method, secondary water resistant barrier if shingles blow off, shutter protection over windows, hurricane straps or hurricane clips, full or partial impact resistant glazing on windows, and other wind resistant construction methods.

### 19.7.2   Against Tornado Damage/Destruction

Tornados are an annual problem in the United States midwest, southeast and Gulf Coast regions. They are called cyclones in tropical and sub-tropical regions (southern hemisphere). The high velocity rotating winds develop a violent vortex that can pull off roofs, demolish homes and other structures, and cause interior water damage. Homeowners insurance policies may include windstorm protection but some exclude tornado damage and loss. Others may exclude roof protection and water damage from horizontal tornado driven rain that gets under shingles and invade a home and over time can cause formation of mold. Available tornado or windstorm insurance may require a high deductible payment. Tornado insurance may or may not be available in high risk areas where there is a history of recurrence and where insurance companies have previously suffered substantial monetary losses. The Texas Windstorm Insurance Association must write policies against tornado damage/loss if other companies refuse. It is important to read a homeowners policy to determine if it covers tornados under "*named perils*" or "*all risks*" clauses. In any policy it may be in a homeowner's interest to have coverage for "all risks" (including tornados), to insure a home to actual value, and be sure that personal possessions (wise to have inventory and photos of the possessions) and living expenses are covered when a home is not habitable while repairs or rebuilding are taking place.

## 19.8    Pollution/Environmental/Workers Compensation Liability Insurance

### 19.8.1    For Environmental Reclamation

This type of policy is individual according to the needs of the insured. It covers costs related to pollution such as cleanup of brown fields and their restoration and injuries or deaths caused by pollution where pollutants contaminated other properties. It also assures all necessary funds for off-site third party property clean-up and victim compensation in the event the polluting entity files for bankruptcy. Businesses or industries that take out environmental liability insurance include contractors, dry cleaners, chemical companies, petroleum and petrochemical companies, manufacturing companies, mining companies, agribusiness, waste disposal companies, and laboratories and medical facilities.

### 19.8.2    To Protect Workers Against Harm from Workplace Accidents

Businesses in many countries are required by law to carry insurance for work related injuries and illnesses. This pays for approved medical, hospital, and related services essential to the injured worker's treatment and recovery. The insurance also provides partial wage replacement for injured workers that are temporarily unable to work. Businesses may also carry disability insurance for workers unable to work again. It should also be the coverage for workers hurt while on the job when a hazard (e.g., earthquake, extreme weather condition, infectious disease) impacts the workplace. This mitigates mental stress to a good degree for the injured or sick workers and their families.

## 19.9    War/Conflict/Hijacking (Piracy) Insurance

### 19.9.1    For Protection Against Property Loss

For the purposes of this section, war/conflict is considered an anthropogenic hazard. As previously noted, homeowners insurance and automobile insurance specifically do not cover losses when there is an act of war such as invasion, revolution, military coup, insurrection, strikes/riots or terrorism if terrorism is deemed to be state sponsored by the State Department. However, if a terrorist act is not shown to be state sponsored but rather carried out by a religious fanatic, fanatic group, or deranged individuals, homeowners policies should cover damage caused by any resulting explosion or fire.

### 19.9.2    Protection for the Transport Sector

For the transport sector, war/conflict/piracy insurance policies are available for ship owners, cargo owners, and airlines. War risk liability covers people or items in the craft whereas the war risk hull covers the craft itself. Premiums, exclusions, or other policy parameters depend on the risk involved where the coverage is to be in force. Businesses may be able to buy separate war insurance policies to cover specific losses.

## References

1.  Marulanda, M. C., Cardona, O. D., Mora, M. G., & Barbat, A. H. (2014). Design and implementation of a voluntary collective earthquake insurance policy to cover low-income homeowners in a developing country. *Natural Hazards, 74*, 2071–2088.
2.  Tian, L., & Yao, P. (2015). Preference for earthquake insurance in rural China: Factors influencing individuals' willingness to pay. *Natural Hazards, 79*, 93–110.
3.  Bielza, M., Conte, C., Dittman, C., Gallego, J., & Stroblmair, J. (2008). Agricultural Insurance Scheme. European Commission, JRC Ispra, Italy, Institute for the Protection and Security of Citizens, 320 pp.
4.  Iturrioz, R. (2009). *Agriculture insurance*. Washington, DC: World Bank. 29 pp.
5.  Mahul, O., & Stutley, C. J. (2010). *Government support to agricultural insurance: Challenges and options for developing countries*. Washington, DC: The World Bank. 207 pp.
6.  Iturrioz, R., & Arias, D. (2010). *Agricultural insurance in Latin America: Developing the market*. Washington, DC: World Bank. 128 pp.
7.  Stutley, C. (2011). *Agricultural insurance in Asia and the Pacific Region*. Bankok: United Nations Food and Agricultural Organization. 225 pp.
8.  Wang, M., Shi, P., Ye, T., Liu, M., & Zhou, M. (2011). Agricultural insurance in China: History, experience, and lessons learned. *International Journal of Disaster Risk Science, 2*, 10–22.

# Chapter 20
# Conflicts/Wars: Human-Driven Events That Injure/Kill People and Damage/Destroy Property

## 20.1 Extent of Conflict as an Anthropogenic Hazard

The world in 2014 was exposed to many conflicts/insurrections/wars (37) that continue in 2016 with people being displaced, injured and killed, property and infrastructure damaged and destroyed. Cumulatively these wars/conflicts and terrorism that accompanies them are at a scale of major natural and/or anthropogenic hazards that kill thousands, injure tens of thousands, and displace hundreds of thousands to millions from their homes or from their nations. This is happening around the globe such as in Ukraine in Eastern Europe; in Libya, Nigeria, Mali, and Eastern Democratic Republic of Congo in Africa; in Afghanistan in Asia; and in Iraq, Syria, and Yemen, in the Middle East, to mention a few. Conflicts/wars kill civilians used as human shields and listed as collateral deaths in on the ground combat or by rockets, artillery shells, and drone/aerial bombing. This is in addition to the deaths of combatants and suicide bombers and their victims. Mitigation can only be found in peace but at this time on our planet, peace does not seem within reach for many of these "anthropogenic" hazards driven by politics, economics, inbred hate, and some in the name of ethnicity or religion.

Cyber warfare falls into this category because it could bring threatening conditions to human populations. As already demonstrated, there have been cyber attacks that have temporarily slowed nuclear processes in Iran (Stuxnet virus), industries have been raided for confidential information, and government computer systems have been penetrated for obtention of classified information. In response, cyber defenses have been created to prevent the introduction of viruses or to prevent "hacking". Are local-regional national governments, industries, infrastructure, and citizens able to withstand an attack on the electrical grid that supports critical systems and if so, for how long? Hospitals and many other critical facilities have generators that react immediately to a power outage. How about urban transport systems without emergency generators such as electric-based trains, subways, and busses that are stopped in place? How many gasoline stations have back up generators to

© Springer International Publishing Switzerland 2016
F.R. Siegel, *Mitigation of Dangers from Natural and Anthropogenic Hazards*,
SpringerBriefs in Environmental Science, DOI 10.1007/978-3-319-38875-5_20

power their pumps? Few if any. Home owners may have generators with a few days of fuel supply but if they need additional fuel, where will it come from given unpowered gasoline stations. Are water treatment plants ready with backup generators and if so, how long will the fuel supplies last? In rural areas, the cut off of electrical power from dairy farms milking machines can be a grave problem unless back up generators are available. How susceptible are communications systems to loss of electrical power or other cyber attacks on critical infrastructure elements and EWSs. These are only a few of the computer/internet problems that could develop unless there is preparedness to prevent, disrupt, or greatly mitigate such impacts with computer defenses and/or other planned ready responses.

## 20.2   Population Displacement, A Human Disaster

The population displacement is a major disaster. By 2014–2016, wars/conflicts have displaced 60 million people from their homes with the great majority trying to find safety within their native countries but with 4–6 million people escaping to other countries where they want to make new homes if they have funds to do so or find refuge in camps (e.g., in Lebanon, Jordan, Turkey). Great numbers try to migrate to countries in southern Europe (e.g., Italy, Greece) or elsewhere in Europe (e.g., Hungary, Croatia, Slovenia) from which they hope to enter "rich" Western European countries (e.g., Germany, Austria, France, England, the Netherlands, Sweden) where they expect to receive financial support, housing, food, and healthcare until they become proficient in a new language and can find jobs. Among those trying to immigrate are not people escaping conflicts/wars but rather refugees seeking economic opportunities. Mitigation of this displaced persons disaster is ideally to end or at least greatly reduce conflicts/wars so that displaced persons can return to a safe and orderly society with human rights, an independent judiciary, education for their children, and employment opportunities. This would seem to be a far in the future possibility given the existing global conditions.

# Chapter 21
# Uncommon but Noteworthy Natural Hazards

## 21.1 Death by Asphyxiation from Lake Eruption of a $CO_2$ Laden Cloud

Lake Nyos is a crater lake in Cameroon. The volcano that underlies it is continually emitting $CO_2$ into the bottom waters. The lake is stratified and the layer of lake water overlying the bottom water traps the $CO_2$ thereby allowing a build up of the gas. In the past, the lake has overturned seasonally releasing the $CO_2$ into the atmosphere relieving the pressure built up during the year. However, in the time preceding 1986, there was no overturn so that the $CO_2$ in the bottom water increased greatly. During the early morning of August 21, there was an event that allowed the eruption of an enormous volume of $CO_2$ into the atmosphere to the degree that more than 15 % of the atmosphere was $CO_2$. An air/$CO_2$ cloud burst forth over the rim of the crater lake and because it was heavier than natural air it flowed down slope into a valley asphyxiating more than 1750 people and thousands of heads of livestock before the cloud dissipated [1]. Where the atmosphere had more than 15 % $CO_2$ there was death. Where there was less than 15 % $CO_2$ in the air because of some dissipation, many survived. What triggered the rupture of the water layer overlying the charged bottom water is unknown but has been attributed by some to a rockfall, by other to a landslide, and by others to a cold rain and the sudden downward current of the heavier cold water through the lighter warmer water. This "eruption" is a rare happening but not unknown. Two years previously at crater Lake Monoun, much smaller than Lake Nyos, 37 people died from $CO_2$ asphyxiation. This was attributed incorrectly by the government to terrorism and someone throwing chemicals into the lake. We know better now. Other crater lakes in Cameroon and elsewhere that fail to overturn seasonally are a natural hazard threat to people and livestock in the nearby vicinity. For example, Lake Kivu is located between Rwanda and Congo, is twice as deep as Lake Nyos, can store more $CO_2$ gas in its bottom water layer from magma leak and bacterial activity, and has two million people living nearby.

© Springer International Publishing Switzerland 2016
F.R. Siegel, *Mitigation of Dangers from Natural and Anthropogenic Hazards*,
SpringerBriefs in Environmental Science, DOI 10.1007/978-3-319-38875-5_21

### 21.1.1   Mitigating the Threat of a Future Occurrence

Carbon dioxide ($CO_2$) monitors with warning systems have been installed at the Cameroon lakes and at Lake Kivu. At Nyos, a possible future problem was solved was solved in 1991 by putting a plastic pipe, perforated at the bottom, into the $CO_2$ bearing layer allowing the $CO_2$ to flow up and disperse in the atmosphere in a slow controlled manner. This is a solution my class in Geological Hazards in Land Use Planning proposed in 1987, and should be the modus operandi for other potentially threatening crater lakes.

It should be noted that a temporal $CO_2$ laced atmosphere can accumulate and exist for a few feet above ground level in the early morning in a volcano crater or low spots on volcano flanks before wind disperses the killer atmosphere. Small animals and birds that come to drink from water puddles on the crater floor have been trapped in seconds and quickly died trying to escape the $CO_2$ enriched atmosphere.

## 21.2   Seiche: Infrequent Coastal Hazard at Lakes or Partially Enclosed Bays

A seiche is a standing wave that develops in an enclosed or partially enclosed body of water (e.g., a lake or a bay) when a prevailing (directional) wind drives water waves onto a shore and they reflect back from there towards the opposite shore. When a reflected wave coincides (resonates) with an incoming wave, there is a harmonic reaction. This causes a rise in the wave crest (node) and a flattening of the background (anti-node). The rise may be only an inch (a few cm) and not perceptible or may be 3–6 ft (~1–2 m) or higher with a drop of the same magnitude when the waves recede and slosh back to the opposite side of a water body, similar to what happens with tsunami secondary waves. A rare seiche that measured 16 ft (~5 m) impacted shore areas of Lake Eire in the United States. With the rise of a seiche wave, there can be local flooding with the rise and damage at shoreline recreational areas (e.g., at marinas), homes, businesses, and infrastructure. The wave motion can be generated during an earthquake as well, not only close to the epicenter, but also thousands of miles away during a major earthquake. Potential problems can be mitigated if deemed necessary, by determining the run up depth of past seiches and zoning the shore area accordingly.

## 21.3   Occasional Food Security Disaster: Desert Locust Swarms in a Regional Swath

Food crops are at risk from desert locusts in northern Africa (e.g., the Sahel), the Middle East, Asia, and southern Europe. The desert locusts act as individuals until breeding season when they emerge as hoppers (young desert locusts) and are forced into small areas with favorable weather and ecological conditions (25 mm [1 in.] of rain each month for two consecutive months and temperature >20 °C [68 °F]) and then begin acting as a group [2]. Within a few months of acting as a group, desert locusts form enormous swarms of millions of locusts that can that can cover many square kilometers downwind, be more than 70 km (43 miles) long, and that travel up to 200 km (124 miles) a day consuming all the vegetation in their path. This includes edible crops and thus affects food security and the livelihood of farmers in fragile ecosystems. During desert locust plagues, the insects may cover or extend into more than 60 countries. Desert locust swarms immediately after or during a drought can significantly reduce food crops as happened in Sahel nations during 2004. For example in Mauritania during 2004, 79 % of cereal crops were lost to the voracious insects. A metric ton of locusts, only a small part of an average swarm, consumes the same quantity of food crops daily as 2500 people and a 1 km$^2$ swarm contains ~40 million locusts that can eat the same amount of food in a day as 35,000 people [3]. To protect its economic interests, Morocco spent US$30 million to safeguard its agricultural sector against desert locusts because agriculture yields US$7 billion of product that includes US$1 billion in export earnings [4].

### 21.3.1   Mitigation Programs

Weather stations throughout the affected regions under the aegis of the World Agro-Meteorological Information Service (WAMIS at www.wamis.org) monitor weather conditions (e.g., rainfall, temperature, wind), satellite imagery, and register direct observations in order to locate breeding areas. Specialists strive to control the emergence and or groupings of hoppers using three methods. One is by spreading poisoned bait such as insecticide laced bran in the paths of the swarms soon after the emergence as hoppers. When the hoppers feed or touch insecticide (organophosphate chemicals), they die. A second is by spraying insecticides on swarms of hoppers or settled adults from vehicles or aerially. The third is by spraying insecticides on vegetation downwind in the path of the hoppers but this latter may be harmful to people or other life forms. For desert locust control that keep food crops useable, farmers need newer, safer insecticides [2, 5].

### References

1. Nasr, S. L. (2009). How did Lake Nyos suddenly kill 1700 people? How stuff works.com.
2. World Meteorological Organization and Food and Agricultural Organization (2006). Regional Workshop for the Anglophone Countries on Meteorological Information Locust Monitoring and Control, Oman, 14 pp.
3. Food and Agricultural Organization of the United Nations (FAO) (2015). Frequently asked questions about desert locusts. www.fao.org/ag/locusts/en/info/faq/. Accessed 2015.

4. Food and Agricultural Organization of the United Nations. (2004). *FAO/WFP crop and food supply assessment mission to Mauritania with special focus on losses due to the desert locust.* Rome: World Food Programme. 2 pp.
5. Mackean, D. G. (2002). *GCSE biology* (3rd ed.). UK: Hodder Murray. 384 pp.

# Epilogue

There are complex problems facing the application of mitigation methods especially for lower- and middle-income economies. One is the funding of projects that lessen the incidence of death and injury for populations and of damage and destruction of property. This may come in part from a nation itself, in part from multinational business interests to protect their programs and as a good will gesture, and in part from donor nations and NGOs, with major input as long term low interest loans from international organizations such as the World Bank or regional banks. The latter three funding sources will require financial transparency, accountability and a stringent control against "major" corruption that can siphon off an important amount of the funding received for hazard impact mitigation. Transparency International evaluated how corrupt perceived their public sector to be in a 2014 listing of 175 countries using a Corruption Perception Index (CPI) [1]. No countries were 100 % free of corruption. Those perceived to be most free of corruption were developed, industrialized economies. Of the 175 countries evaluated, 120 fell below the CPI 50 and were for the most part less developed and developing nations, suggesting that to a greater or lesser degree, corruption holds back national development. Because of the degree of corruption in many countries that need this funding, it may not be available or not given in the amount necessary to carry out needed mitigation projects without stringent financial oversight.

A second problem is implementation of a project with people who have experience in the design, building, operation, and maintenance in the technology or non-technological methods being used in the very different mitigation methods described in the text. For example design of dams requires a complete detailed geological study as described in the section on dams (e.g., landslide possibilities, reservoir induced earthquakes). It requires an assessment by sociologists and biologists as to displacement of citizens and possible loss of arable land plus a study on ecosystem disruption by rising waters behind the structure as well of downstream ecosystems

© Springer International Publishing Switzerland 2016
F.R. Siegel, *Mitigation of Dangers from Natural and Anthropogenic Hazards*,
SpringerBriefs in Environmental Science, DOI 10.1007/978-3-319-38875-5

that lose the normal flow of river waters and sediment deposition. Without experience learned over time in like projects, decision makers and construction managers and engineers are doomed to repeat errors that were made in the past but that others learned from and corrected in subsequent like projects. Similarly design and construction or retrofitting of structurally endangered buildings in earthquake prone areas by technically knowledgeable personnel is absolutely necessary. Essential facets of mitigation programs such as these work for the methodologies applied to reduce disaster situations from other hazards.

A third problem is the need to plan against the progressive nature of global warming/climate change and the hazards they fuel that were discussed earlier in the text. For example, this is especially important for the 200 million people today living in coastal areas less than 5 m above sea level that are threatened by sea level rise as would be the estimated 400–500 million living there by the end of the century, mainly in Asia with its coastal mega-cities. Large percentages of population in Bangladesh and Viet Nam live in low lying coastal regions and are especially vulnerable to extreme weather events [2]. The danger comes mainly from high energy storms (hurricanes, typhoons, monsoons), storm surges, and flooding. The same vulnerability concerns that are intensified or expanded can be applied to populations exposed to dangerous flooding today and in the future, and to the spread of infectious diseases.

The need to implement mitigation processes now in 2016 and maintain it as a continuing program from year to year is emphasized by two population predictions and the status of existing sustainability conditions we reiterate here (Table 1.1). First are the previously cited projections that the global population is expected to rise from 7.3 billion people in 2015 to 9.8 billion by 2050 and perhaps stabilize at 11 billion or more by the end of this century [3]. There is the real question as to whether we can sustain added population, for example, with water, food, safe shelter, medical care, and employment to reduce poverty and disaffection and despair especially by the youth in these growing populations. This is stated in light of the facts that in 2015 one in nine people on earth were malnourished, one in seven were without access to safe water, and more than two billion did not have access to adequate sanitation or electricity as starting points. Will these numbers grow as populations increase in the future or can technological advances assure sustainable conditions for all? Second, in addition to the global population increase, there will be demographic changes from natural urban births and as more people move from rural settings to urban centers and immigrants settle in cities. As noted in Chap. 2, In 2015, 3.7 billion people were urban dwellers and 3.6 billion lived in rural areas. By 2050, there is projected to be a huge change with 6.9 billion people living in cities and 2.9 billion in rural regions. This brings an increase in population density to urban centers. In 2015 there were more than 500 cities with more than one million inhabitants world wide but especially in Asia [4]. There were 34 mega-cities (>10 million inhabitants) in 2015 with the nine of the twelve most populated being in Asia (Table 1.2). This number will continue to increase. Mitigation of natural and anthropogenic hazards that might prevent them from becoming large scale disasters has to be a national priority now as you read this book, especially where there is a high density of population but also where there are lesser numbers of people that are exposed. Only this can lessen deaths and injures to populations and reduce damage and destruction of structures and infrastructural elements so as to keep societies functioning. It has been shown over and over again, in theory and in practice, that it is far less costly to apply mitigation technologies before hazards impact than to absorb the costs to respond to a disaster with the necessities of life and search and rescue or search and recovery efforts and then move onto the post event rehabilitation and reconstruction phases.

# References

1. Transparency International. (2015). *Corruption perception index 2014*. Berlin, Germany, 8 pp. View www.transparency.org/2014/results
2. World Ocean Review. (2015). Living with the oceans. A report on the status of the world oceans. Living in Coastal Areas. Accessed 2015, from www.world-oceanreview/en/wor-1/coastal/living-in-coastal-areas
3. Population Reference Bureau. (2015). *World population data sheet*. Washington, DC, 21 pp.
4. Demographia World Urban Atlas. (2015). 11th annual edition, 2015:1 134 pp. www.demographia.com/db-worldua.pdf

# Index

## A
Acid mine drainage (AMD), 17, 93
Acid rock drainage, 93
Active Mass Driver, 25
Africa, 4, 69–71
Agricultural (crop/livestock) insurance, 107–108
Air pollution
  indoor, 79–80
  outdoor, 80–81
All-or-nothing (AoN), 45
Anthropogenic hazard, 113–114
Arsenic poisoning, 81
Asia, 4, 13, 45
Asthenosphere, 19
Automobile coverage, 103
Avalanche, 39, 40

## B
Benefit/cost relation, 97
Building codes, 24, 25, 29, 32, 95

## C
California, 103
Carbon dioxide ($CO_2$), 116
China, 23
Climate zones, 61
Columbia River, 69
Compaction, 41, 42, 55
Conflicts/wars
  anthropogenic hazard, 113–114
  population displacement,
    human disaster, 114
Cost effectiveness analysis, 98

## D
Dam placement, 76
Dam siting
  failures/disasters, 75–76
  geology, 74
  human/ecosystems effects, 76–77
  hydroelectric dams, 74
Democratic Republic of Congo (DRC), 36
Desalination, 57
Desert locusts, 116
Desertification
  definition, 63
  drylands, 63
  global warming/climate change, 64
  socio-economic factors, 64
Development corridors, 70
Diarrheal disease, 48
Disease
  humans, vectors, 62
  latitude and altitude, 61
Disease vectors, 62
Drainage basin, 73
Drainage basin monitoring, 35
Drought, 56, 57
Drought insurance
  crop/livestock loss protection, 108
  heat waves, 109

## E
Early warning systems (EWSs)
  earthquakes, 15, 16
  extreme weather conditions, 16
  hazard monitoring, 17
  infectious disease, 16

© Springer International Publishing Switzerland 2016
F.R. Siegel, *Mitigation of Dangers from Natural and Anthropogenic Hazards*,
SpringerBriefs in Environmental Science, DOI 10.1007/978-3-319-38875-5